Simple Earth & Space Science Investigations

Solutions for doing science in the classroom.

Christopher P. Garside

Seven Sides Publishing

Seven Sides Publishing has a mission to improve the teaching and understanding of science. To contact us, send an email to simpleinvestigations@sevensidespublishing.com or visit our website at sevensidespublishing.com.

Copyright © 2022 by Seven Sides Publishing and Christopher P. Garside. All rights reserved. No part of this publication may be reproduced, stored in a retrieval system, scanned, or transmitted in any form or by any means, electronic, mechanical, photocopying, recording, or otherwise, without the prior written permission of Seven Sides Publishing and Christopher P. Garside. Photocopying is permitted from this book to make copies only for the students of the teacher who owns this book; this is not for other teachers, students, or anyone else in the school or district's students.

ISBN: 9798420930588

Published by: Seven Sides Publishing, Cypress, TX.

Table of Contents

Introduction	Page 4
Unit 1 The Development of the Universe *ESS1.A*	Page 17
Unit 2 Characteristics of the Solar System *ESS1.BC*	Page 43
Unit 3 Sun Earth & Moon Interactions *ESS1.B*	Page 86
Unit 4 Earth's Structures & Functions *ESS2.AB*	Page 105
Unit 5 Weather Patterns *ESS2.CD*	Page 141
Unit 6 Energy Flow & Cycles *ESS2.D*	Page 173
Unit 7 Ecological Changes *ESS2.D*	Page 195
Unit 8 Managing Natural Resources *ESS3.AB*	Page 228
Unit 9 Human Impact on the Hydrosphere *ESS3.BC*	Page 255
Unit 10 Influences on Climate Change *ESS3.D*	Page 285
TEKS and NGSS Correlations	Page 306
Equipment List for all Investigations	Page 313

Introduction

To help teachers teach science through investigations, Seven Sides Publishing has provided a series of lab manuals for Elementary Science, Middle School Science, Biology, Chemistry, Physics, Earth and Space Science, and Environmental Systems. These manuals are a rich resource for structure and investigations. There is a shortage of user-friendly labs that easily allow teachers and students to perform experiments quickly. Too many labs have too much busy writing within them, where teachers do not want to take the time to read everything to figure out if it would be good for them to use with their students. If the teachers do not want to read it, do you think the students do? So we have taken a lot of the traditional labs that have been around for decades and simplified them; so they are easy to read and perform. We have also added some new original labs that have never been seen before. There have been efforts to have teachers do more investigations with their students, but there is no plan or solution to deal with the real issues teachers have in preparing to do this. The book How to Teach Science Through Investigations has the plan, and the Simple Investigations Lab manuals have the solutions so students can learn science through investigations with minimal effort. Teaching science through investigations will make your classrooms more efficient, where students learn content and practice skills simultaneously. Science is a process of doing. Doing this process is the most efficient way for students to learn science and be able to use it in the future. We live in a culture where science-literate people are needed for jobs, but too few can be found. If you incorporate these labs with virtual labs (that I will point you to in each section of the lab manual), skill/math practice, and concept maps, you will not need to fill in gaps by giving lectures. All content can be learned through investigations and practice. Remember, we only remember 5-20% of what we hear. That 20% is when you are really interested in the content. But hearing practices no science process skills and does not activate any higher cognitive thought. Lecturing is not a good option. We remember 75-80% of what we do/experience and 90-95% of what we teach. Investigations allow us to keep our students in these higher retention percentages. Teaching through investigations also works because students spend more time in class at higher Bloom's Taxonomy levels, staying in zones C and D on the Rigor Relevance Chart when they perform investigations. And if you add the physical way they are stimulated with the hands-on experience, you cannot deny the level of learning will be much higher while students perform investigations. This manual gives you the resources you need to teach Earth and Space Science through investigations.

We separated each of these sections in the manual like you may divide your class units. We will follow the Next Generation Science Standards to make it easy for you to find the labs you want and need for your classes. We include concept maps at the front of each section that shows the vocabulary and visual clues to how concepts relate to each other; this is a great way to organize information. It talks to the students to see how ideas work together, making it easy to chunk information to use at higher cognitive levels. At the beginning of each lab, we put the materials you will need in boldface in the directions; this saves time for your lab preparation. There is also a safety question in boldface just after that for you and your students to evaluate. It says, "Looking at the material and lab we will be using, what are the safety precautions we should take to protect ourselves and materials during this investigation." Make sure to read the lab so you can better answer this question with your students.

Virtual Labs

Hands-on labs are not the only way for students to learn science, but they are the most effective. However, virtual labs can be used with these hands-on labs. Many investigations physically cannot be done hands-on, so some experiments will have to be done virtually. There are three sources that I have used in the past that have a good number of resources. **Physicsclassroom.com** and **PhET.colorado.edu** are free to everyone and are great to use. **Physicsclassroom.com** has teacher notes and activities/exercises that guide students through Physics and Chemistry Interactives. You can find them under the simulation and open, download, or print the PDF. They also have a series of Concept Builders that are a tremendous virtual practice that can replace worksheets that help students practice concepts, math, and skills. They can be hard to find, so above the list provided is the section where they can be found (underlined and in italics) on the website. **PhET.colorado.edu** has a variety of activities of different levels that you can explore to go through their simulations. They are also easy to download and print. **ExploreLearning.com** is expensive, but the quality of its product is much higher than the other two. When you click on a Gizmo, you can also click on lessons and find the Student Explorations that go with each Gizmo that you can modify, download, and print. They are written at a very high quality, making the students think like a scientist. At the end of each section of this lab manual, we include a list of virtual labs from these organizations that would be great to use with these labs. Please remember virtual labs should never replace hands-on labs. If the students can learn the content live, that should be the priority because it is more of an experience that will be remembered. There are many other virtual simulations out there, but none so far have moved me to use them over the three I have mentioned here.

Probe-ware

This lab manual has lots of labs that use probe-ware. Students must learn how to use probe-ware; this means teachers need to know how to use probe-ware. Many companies use digital probe-ware with all the research, development, testing, and forensic testing they do;

this has potential career opportunities that help students become more marketable for jobs if they are familiar with using probe-ware. Hooking everything up is just as easy as charging your phone. When I was a High School Science Technology Coach and researched which companies and devices would be the most user-friendly to students, I found using Vernier Probe-ware was better for high school students, but PASCO seemed better for middle school students. Both are giants in the probe-ware industry for education. Since I am more familiar with Vernier, I will be referring to Vernier Probe-ware. However, PASCO would be a great alternative.

Interfaces are devices that the probes are connected to that talk with the program (Logger Pro) that displays the data. I found the most economical and friendliest way for students to see the data from probe-ware is to use the Vernier LabQuest Mini interface hooked up to a computer with Logger Pro. LabQuest Mini has multiple ports that are needed in many labs. They are the least expensive, so they are better on the budget. They require no batteries, so they are easy to transport if you need or want to. The other interfaces are more expensive, require batteries if you are going outside, or the stand-alone devices have a smaller screen to see the data, with less flexibility to manipulate the parameters like changing the time of data collection or changing units if you want to change or modify an experiment. Some costly wireless probes and interfaces may be easier to use if you do not mind the cost. A computer screen is much bigger and makes it easier to see the data, so this is my preferred setup. But using any interfaces will work fine for these labs.

Connecting the Probe-ware

To hook them up, you will plug your probe into one of the channels or the sonic on the interface. If the plug does not fit in smoothly, either you are plugging it upside-down or trying the wrong port. Then take the little chord that looks like it would go into your phone and plug that into your interface. Take the other end, and plug it into a USB port on your computer. Open up Logger Pro on your computer. If everything is hooked up properly and the computer and interface are working properly, you will see a green button at the top of the computer screen that says "Collect." Many of the labs have preset settings in Logger Pro. You will use the manila folder at the top left of the toolbar in Logger Pro to find the folders and files you will be instructed to go to for these specific settings for different labs. Whenever you get the physical equipment, they will have detailed instructions in the box they come in on how to hook them up if you are still confused. They will also have instructions on how to calibrate the probes if needed. There are a few probes that require frequent calibration. If we use any, it will be discussed in the lab directions. The more you use probe-ware, the easier it gets to set up. I usually only have to show my students twice to have them be able to set the equipment up on their own. But as you are showing them, have them physically do it. You can also find detailed instructions online at Vernier.com. Many more detailed labs can also be found there under lab ideas.

You also can use standard equipment like spring scales for force sensors or thermometers for temperature probes. Because schools want to integrate more technology, we wrote these labs to use probe-ware wherever applicable. Because they are so simple, these labs can be modified to fit whatever equipment you have. There are very few labs that I have used in my career that I did not alter how I presented them. One reason we wrote these labs this way is to customize them to the Texas TEKS and National Standards. We also wrote them how we thought a teacher would want to use them.

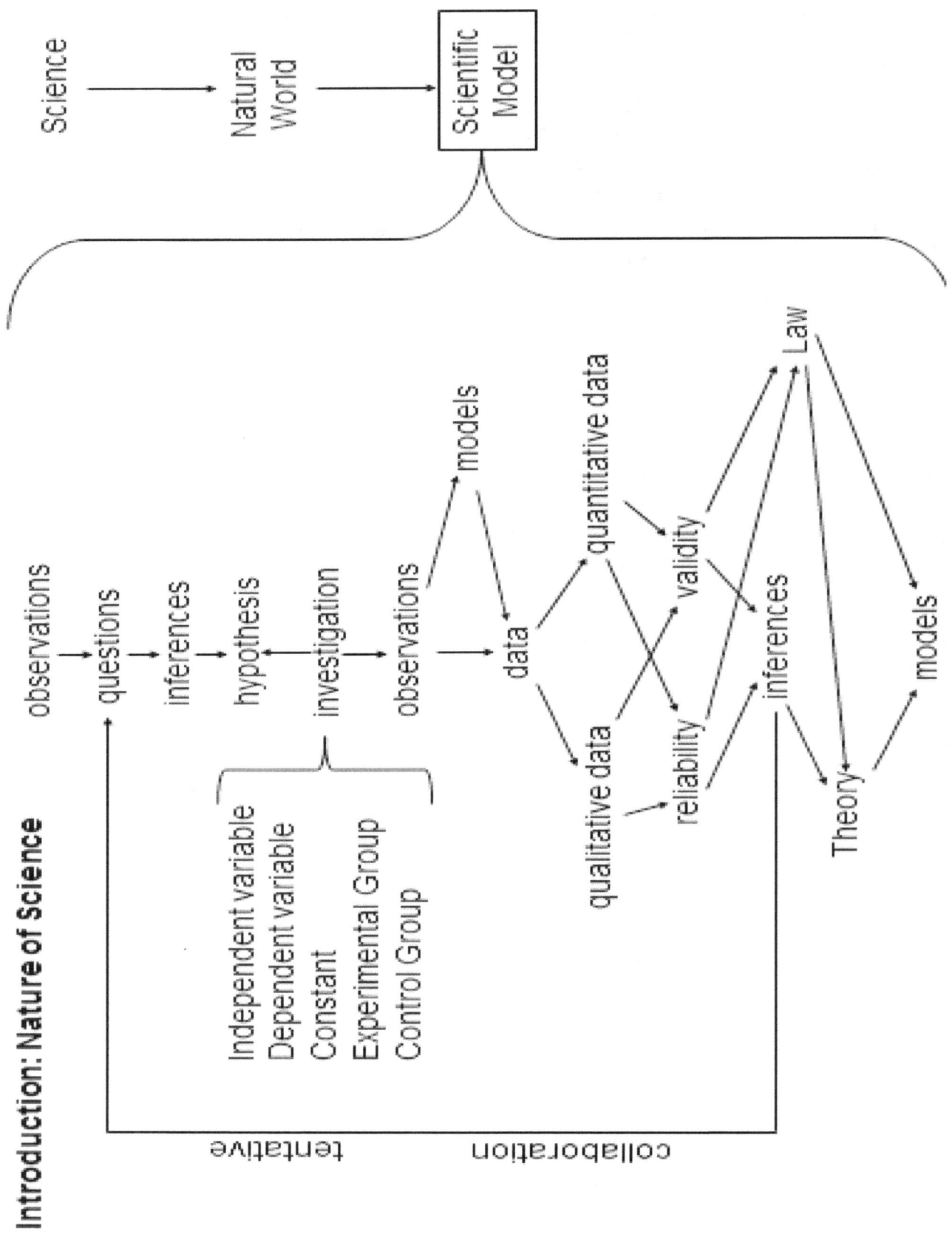

Focus on the Process

Directions:

Get a **small Legos set**. Teachers, make sure it is not too easy for your students. You are going to try to put it together in two different ways. Time how long it takes to put it together each way and answer the questions that follow. **Looking at the materials and lab we will be using, what are the safety precautions we should take to protect ourselves and materials during the investigation?**

   A) Take the Lego pieces and construct the picture (the **product**) on the box's cover, looking at nothing but the cover and the Lego pieces.

   B) When 20 minutes have passed, or you are done, take what you have made totally apart. Take out the directions (the **process**) and construct the product while using the step-by-step directions. Time how long it took you to complete the set.

Questions:

   1) How did it feel trying to construct the Legos (A) without any directions?

   2) Did you finish? If so, how long did it take?

   3) How did it feel to construct the Legos (B) with step-by-step directions?

   4) Did you finish? If so, how long did it take?

   5) Which strategy (A or B) allowed you to complete the product?

   6) Which strategy (A or B) was more intimidating?

7) Which strategy (A or B) allowed you to see what is under the surface?

8) Which strategy (A or B) will allow you to learn more?

We often get anxious or procrastinate when faced with a large task. We are tempted to take a "shortcut" (copy or cheat, we do not learn much when we do this). There are pain and stress hormones that are released when this happens. One way to overcome this is to just worry about the next step in the process and not worry about the product. You can see and measure progress, which makes the process not feel too bad. Another way is just to start working. When you start working, those pain and stress hormones stop getting released so that anxiety goes away; this is why when we want to learn efficiently and effectively, we must:

Focus on the _____ and the _____ will take care of itself.

9) How is putting the Lego pieces together like putting ideas together to understand concepts?

Measurement Lab

Directions:

You will need **water**, a **scale**, a **meter stick**, a **temperature probe** attached to an **interface** connected to a **computer** with **Logger Pro**, a **100 mL graduated cylinder**, and a **stopwatch**. **Looking at the materials and lab we will be using, what are the safety precautions we should take to protect ourselves and materials during the investigation?**

1) Take the graduated cylinder and find its mass empty; write this in Data Table 1.
2) Add 50 mL of water to the graduated cylinder. Make sure you use the meniscus properly where the volume is at the bottom of the meniscus. Have the teacher check that you measured it correctly. Have each person in your group empty and fill the graduated cylinder with 50 mL of water. As they do so, have each person in your group time how long it takes for each person to fill the graduated cylinder and check it is correct (it is not a race, just a chance to get familiar with using the graduated cylinder and stopwatch).
3) Now find the mass of the graduated cylinder with 50 mL of water in it. Subtract the mass of the empty graduated cylinder from this mass and write the water's mass in Data Table 1.
4) Connect your temperature probe to an interface and connect your interface to a computer with Logger Pro (unless you have a LabQuest 2, then just hook your probe to the LabQuest 2). Find where the Logger Pro is located on your computer so you can use it again in the future. Once open, find the graduated cylinder's water temperature in Fahrenheit and Celsius (you will have to figure out how to change units). Write these in Data Table 1.
5) Take your meter stick and measure the length of the graduated cylinder. And measure the width of the base in centimeters. Write these in Data Table 1

Data Table 1

Object	Mass (g)	Volume (mL)	Time to Fill (s)	Temp (°F)	Temp (°C)	Length (cm)	Width (cm)
Graduated Cylinder		✖		✖	✖		
Water			✖			✖	✖

Questions:

1) Convert a length to meters, the volume to liters and a mass to kilograms, and Celsius to Kelvin.

   Length _____ m   Volume _____ L   Mass _____ kg   Temp _____ K

2) What do you notice about the mass of the water compared to its volume?

3) What can happen to your investigations if your measurements are not accurate or precise?

4) Why do you think the rest of the world uses the metric system over the English system.

Patterns in Pennies

Directions:

You will need a **ruler**, 10 **pennies**, a **scale**, a **roll of pennies**, and an **empty penny roll. Looking at the materials and lab we will be using, what are the safety precautions we should take to protect ourselves and materials during the investigation?**

1) Find the mass of one penny with a scale to the nearest .1 g. Then measure the height of the penny in millimeters. Write these in Data Table 1 below.
2) Place another penny on top of the original penny and find the mass and height of the two pennies. Write these in Data Table 1 below.
3) Keep adding pennies one by one, measuring the mass and height until you have 10 pennies on the scale.
4) Make a line graph with the mass on the (x) axis and the height on the (y) axis for the pennies on Graph 1.

Data Table 1

Number of Pennies	Mass	Height
1		
2		
3		
4		
5		
6		
7		
8		
9		
10		

Graph 1

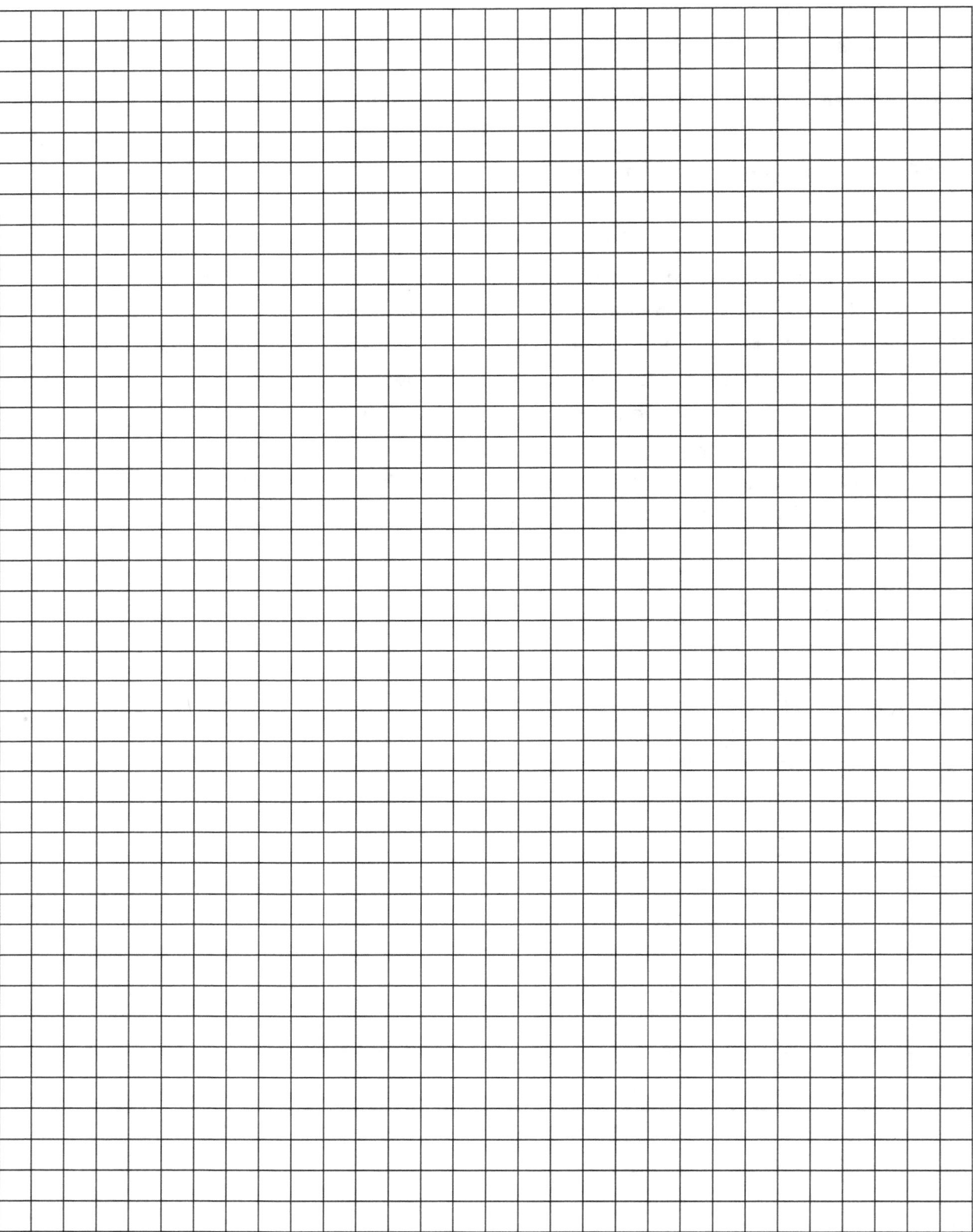

Questions:

1) What do you notice about the graph?

2) Is this a direct or inverse relationship between mass and height?

3) Do all pennies have the same mass? (Explain)

4) Do all the pennies have the same thickness? (Explain)

5) Use your data to estimate how many pennies are in the coin roll. How many pennies do you think are in the roll?

6) What did you do to estimate the number of coins?

7) What else could you do to estimate the coins?

8) Try your answer to #7. Do you get the same number as #5?

9) Carefully open up the coin roll and find out how many pennies there are. How close were you to the real number? After you are done counting, carefully close the roll back up.

10) Calculate the % accuracy by taking the lowest number between your guess and the actual number dividing by the higher of the two, then multiplying by 100.

11) What were some sources of error?

Virtual Investigations that go with Introduction

ExploreLearning.com

    Unit Conversions Gizmo

    Graphing Skills Gizmo

    Measuring Volume Gizmo

    Elevator Operator (Line Graphs) Gizmo

    Weight and Mass Gizmo

    Triple Beam Balance Gizmo

    Reaction Time 1 Gizmo

    Reaction Time 2 Gizmo

Physicsclassroom.com/Concept-Builders/Chemistry:

    Measurement and Numbers

    Significant Digits and Measurements

    Metric System

    Metric Estimation

    Experiments and Variables

    Proportional Reasoning

    Calculating Slope

    Using Graphs

    Which One Doesn't Belong

Unit 1: The Development of the Universe

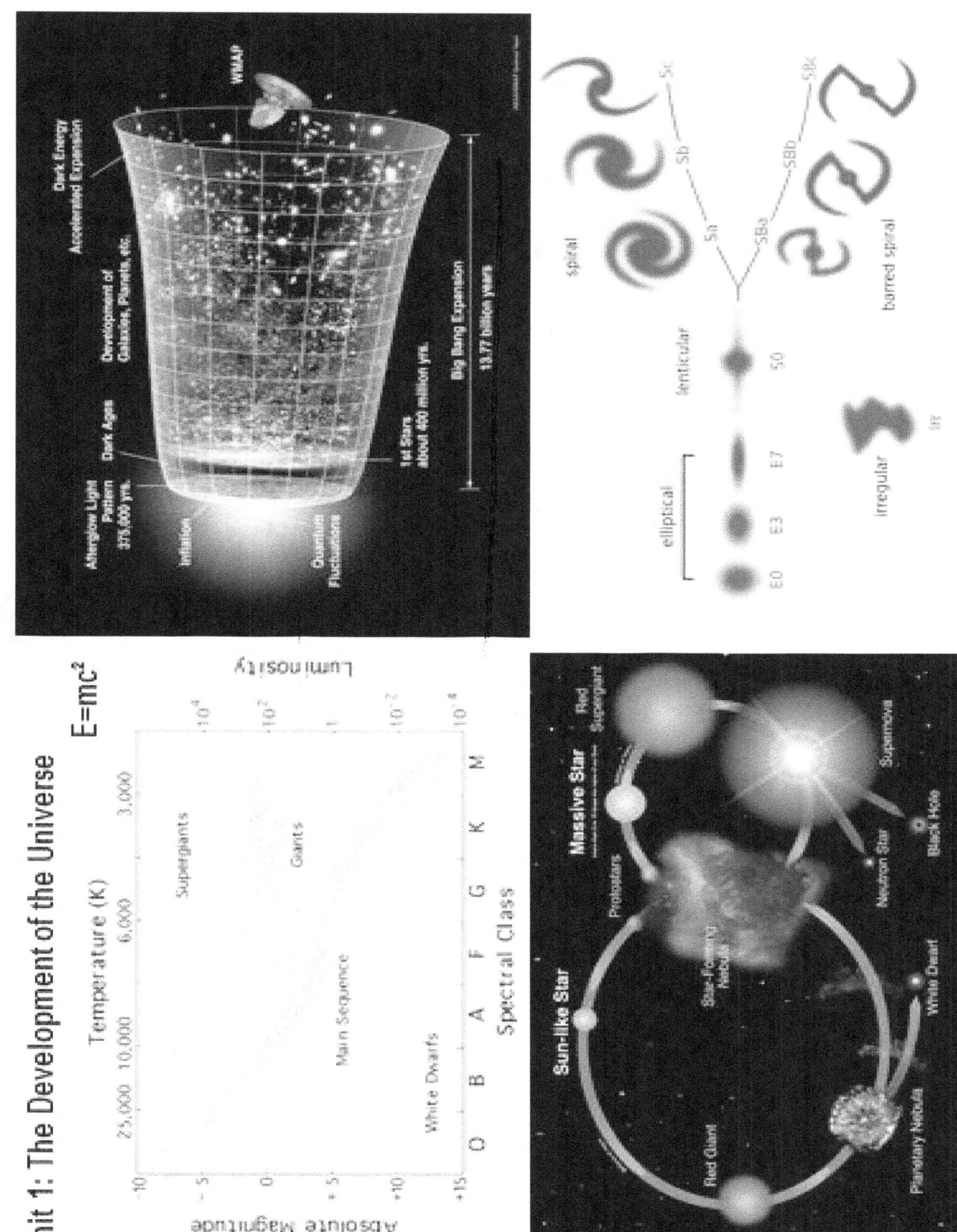

Star's Life Cycle

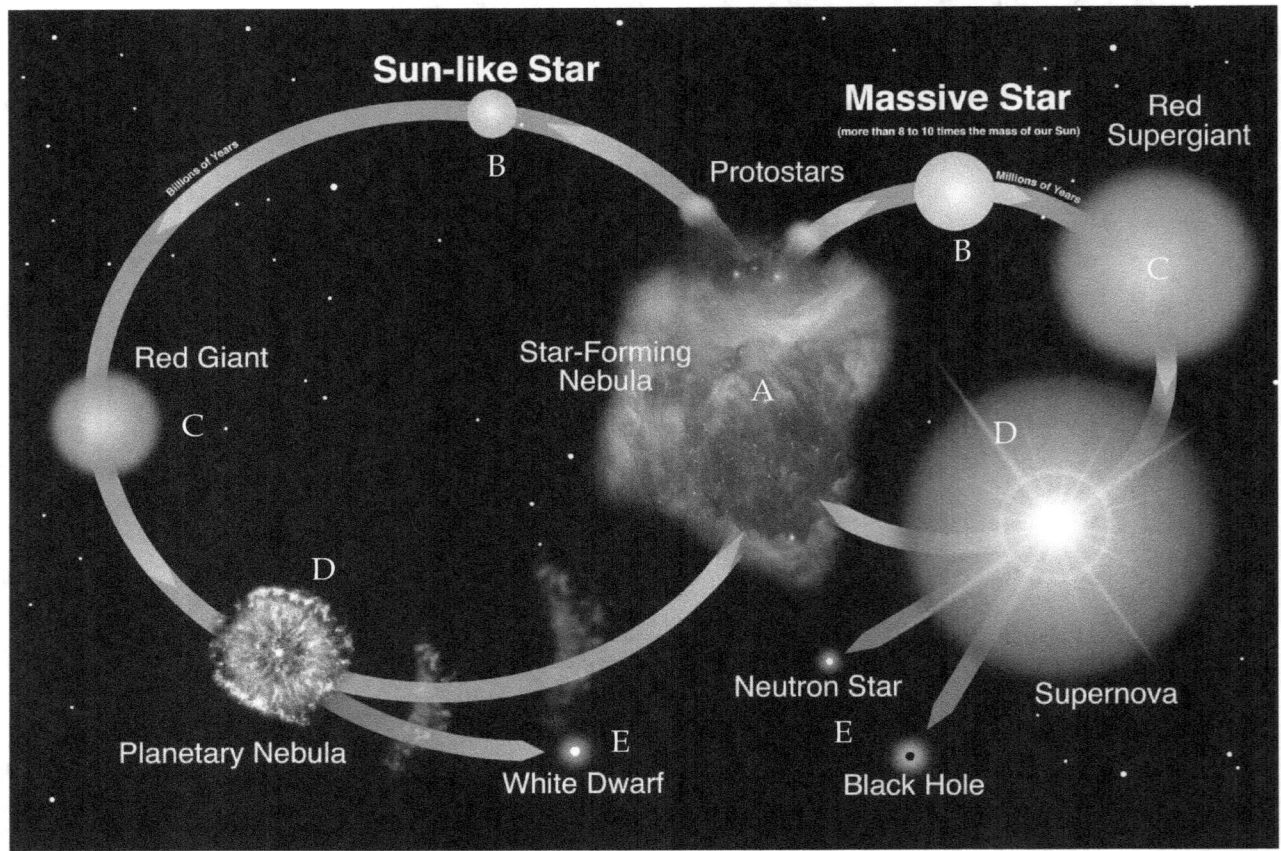

Directions:

Use the diagram above from NASA to answer questions about the life cycle of stars.

1) What is the difference in how long a star will live between normal stars like our sun and massive stars?

2) How do all stars start their life?

3) How do all stars end their life?

4) Why is this called a cycle?

5) In a star, gravity battles with fission and fusion reactions that allow it to remain a star. Fission and fusion allow it to expand, and gravity makes it collapse. As it ages, it speeds up its fuel burning and starts expanding. What letter on the diagram above are the stars expanding?

6) Then the star runs out of fuel, and its gravity makes it collapse on itself, causing a great explosion. What letter on the diagram shows the star exploding?

7) After it explodes, what will our sun leave behind?

8) After it explodes, what can massive stars leave behind?

9) How do you think an aging sun will affect life on Earth?

10) Do we have to worry about that in our lifetime? Explain.

11) Explain how a large star could give rise to smaller stars.

12) If we want to look for more evolutionarily advanced life, what type of stars should we focus on to look for planets with that kind of life? Explain.

Star Life Cycle Model

Directions:

You will need **round balloons of different sizes**, one for each group. **Looking at the materials and lab we will be using, what are the safety precautions we should take to protect ourselves and materials during this investigation?**

1) The empty balloon is like the shapeless **nebulae of gas** that will form a star.
2) Take a balloon and blow air in it until you get it to start filling up; this will represent when the **star forms**.
3) Look at the different sizes of balloons in the class; this represents the different sizes of stars.
4) As the star ages, it will speed up the burning fuel, and the star expands. Blow air into the balloon showing the **aging of the star**.
5) Eventually, the star will run out of fuel and **explode into a supernova**. Keep blowing air into the balloon until it explodes.
6) As the stars explode, listen to each balloon; the bigger the balloon, the bigger the sound; just like the bigger the star, the bigger the star's explosion.

Questions:

1) How did this model correctly show the lifecycle of stars?

2) What did this model not show about the lifecycle of a star? What was missing?

Our Bright Morning Star: the Sun

Directions:

Use the **internet** to help you answer the questions that follow.

1) What are the ten things we need to know about the sun?

   a.

   b.

   c.

   d.

   e.

   f.

   g.

   h.

i.

j.

2) What is the diameter of the sun?

   a. How does that compare to Earth?

3) What does the sun orbit?

4) When and how did the sun form?

5) Describe the structure of the sun.

6) Describe the surface of the sun.

7) Describe the six key features of the sun.

   a. Sun Spots:

b. Coronal Holes:

c. Solar Flares:

d. CME:

e. Solar Prominence:

f. Spicules:

Nuclear Fission and Fusion in a Star

Directions and Questions:

Use the **internet** and your **textbook** to research and explain how nuclear fission and fusion happen inside a star.

1) Draw a diagram of nuclear fission taking place in the sun.

2) What are the element isotopes involved?

3) How does the reaction get started (what goes into the reaction)?

4) What comes out of the fission reaction?

5) How does this lead to a chain reaction?

6) Draw a diagram of nuclear fusion taking place in the sun.

7) What are the element isotopes involved?

8) How is fusion different than fission?

9) What goes into the fusion reaction?

10) What comes out of the fusion reaction?

11) How do these reactions fight gravity in the sun?

12) What happens to the star if these reactions stop?

Compare and Classify Stars

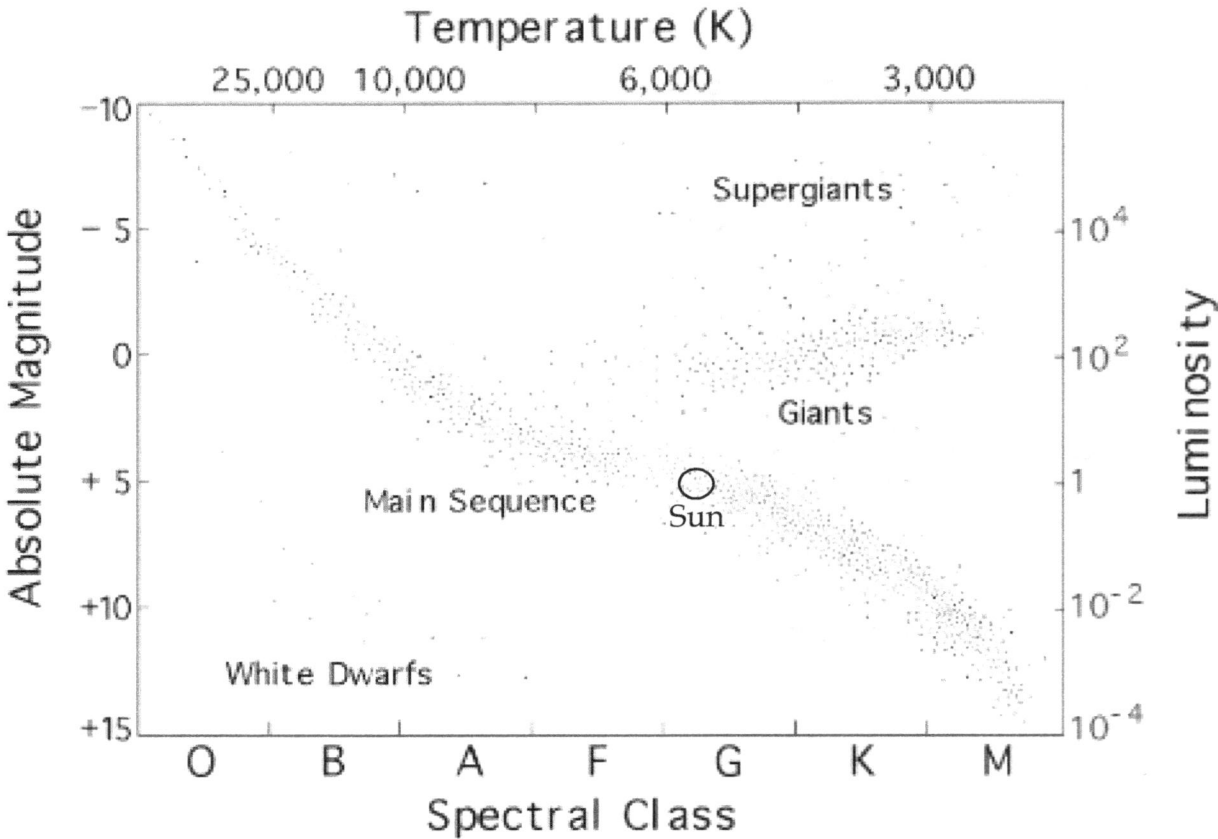

Directions:

Use the Hertzsprung-Russell diagram from NASA above to answer the questions below.

The **surface temperature** is on the top margin, the **class** of the star is on the bottom, absolute magnitude (how bright it would be if viewed from 10 parsecs or 32.6 lightyears away) is on the left margin, and **luminosity** (brightness) is in the right margin.

The most important factor in distinguishing one star from another – temperature, luminosity, size, and life cycle – seems to be the **star's mass**.

1) What classification is our sun?

2) What is our sun's surface temperature?

3) How does the size of the star determine its brightness?

4) How does the size of the star determine its surface temperature?

5) According to the diagram, what seems to determine the class of the star?

6) Does the class of the star determine its brightness? Explain.

7) The sun will one day become a red giant. Will the sun become brighter or darker? Explain how you know.

8) How will this affect life on Earth?

9) More massive stars have a much shorter life cycle (in millions of years) than our sun (10 billion years); in turn, less massive stars, like red dwarfs, have extremely long life cycles (possibly over 100 billion years). Should we expect to find well-evolved life (like on Earth) on a planet orbiting a giant or supergiant star?

10) If we could find it, where should we look for intelligent life outside of Earth (giving it more time to evolve)? Explain why.

Classifying Galaxies

Directions:

Use the **internet** to go to a NASA webpage using the URL: www.tinyurl.com/2p9xvay6. Use this page to answer the questions below.

1) What are the three main types of galaxies? Draw a picture of each.
    a.

    b.

    c.

2) What are the sub-classifications of the spiral galaxies with bulges? Draw a picture of each.
    a.

    b.

    c.

3) How are spiral galaxies with bulges different from ones with bars?

4) Draw pictures of some examples below.
   a. Bulges
   b. Bars

5) What are the sub-classifications of spiral galaxies with bars?
   a.
   b.
   c.

6) Why is there a number range for sub-classifications of elliptical galaxies? Draw pictures of them below.

   a.
   b.
   c.
   d.

7) What is the reason for the last classification of galaxies?

A Guide to the Milky Way Galaxy

Directions:

Use the **internet** to help you answer the following questions.

1) How would you describe where our solar system is located in the Milky Way galaxy?

2) Which arm are we on?

3) What is our spur?

4) Which arms of the galaxy are outside us?

5) Which arms are inside us?

6) Like the earth orbits around the sun in our solar system, our sun orbits a supermassive black hole at the center of our galaxy. The energy created from the friction of its distortion of the space-time continuum appears as a glow (we cannot see it because lots of gas is in the way blocking it) on the map. How many light-years (ly) are we from the center of the supermassive black hole?

7) How wide is the Milky Way?

8) How many stars are found in the Milky Way?

9) How did the scientists map our galaxy?

10) Is this a perfect map? What stops us from seeing our own galaxy?

11) Use the two maps from the NASA webpage to help you draw a map of the Milky Way galaxy on the rest of this page. Make sure to show and label all the important objects we know of. Use a compass to help you draw the circles showing the distance in lightyears from our sun.

Ten Things NASA Wants you to Know about the Universe

Directions:

Use the **internet** to help you answer the questions that follow.

1) What was the big discovery Hubble made?

2) What kind of matter and energy make up the universe? Give percentages of each.

3) What is the universe mostly made of?

4) How many galaxies are in our cosmic neighborhood?

5) Are there more planets or stars in the universe? Explain why.

6) What is the shape of the Milky Way?

   a. What other shaped galaxies are there?

7) Watch the NASA video: Our Milky Way Galaxy: How Big is Space? How does light help us measure the size of the universe?

   a. How far is a light-second?

   b. How far is a light-minute?

   c. How far is a light-hour?

   d. How far is a light-Day?

   e. How far is a light-year?

8) Have we found other life out there?

   a. Do you think we will? Explain why.

9) What is at the center of the galaxy?

10) How many galaxies could be in our universe?

   a. How many were pictured in the Hubble eXtreme Deep Field picture?

The History of the Big Bang Theory

Directions:

Use the **internet** and your **textbook** to research the history of the Big Bang Theory of our universe. Then answer the following questions.

1) How does the Second Law of Thermodynamics show the universe had a beginning?

   a. Draw a graph of usable energy over time, showing the evidence.

2) What does Einstein's Theory of Relativity say about the relationship between matter and energy?

   a. Write his famous equation here.

3) Who used Einstein's equations to show mathematically the universe was expanding? (hint: there are two that are famous for this)

4) Who then saw physical evidence of this with a telescope?

   a. How did he use the Doppler Effect to show the universe is expanding?

5) Describe what the WMAP is and what it shows us.

6) What atom characteristic shows that the whole universe can fit inside a space smaller than an atom?

   a. What other characteristics of the atom show the energy of matter?

7) What are NASA and other scientists currently working on to help us learn more about the Big Bang?

8) What is quantum mechanics, and what does it show about the matter that makes up our universe?

9) What is the problem with Relativity and Quantum Mechanics?

10) Describe three other models that try to explain the origins of our universe.
    a. 1st Model

    b. 2nd Model

c. 3rd Model

11) Which of the models you researched today do you think best describes the origins of our universe? Explain why.

The Pixel of the Universe

Directions and Questions:

You will need a **golf ball**, a **bead**, and a **large field** or **parking lot. Looking at the materials and lab we will be using, what are the safety precautions we should take to protect ourselves and materials during this investigation?**

1) Walk out to a large field or parking lot, at least the size of a football field. Keep in mind the space you use still may be too small to be a scale model. You will make a model of a hydrogen atom with 1 proton and 1 electron.
2) On one edge, take a small red bead representing an electron and put it somewhere where you can see it (hang it on a fence or a tiny branch).
3) Walk at least 100 yards away; if you have more room, you can use that. Hold up the golf ball (a proton), read the information, and answer the following questions. Can you see the bead?

4) This distance is how far away the closest electron speeds around the proton. The speed approaches the speed of light. It moves so fast that it makes a ball the size of a football stadium. If you have ever seen a fan moving fast, does it look like a disk? But is it a disk?

    a. So we have to ask ourselves, the atoms, the pixels of our universe, look like solid balls (see the picture below), but are they balls? Explain why.

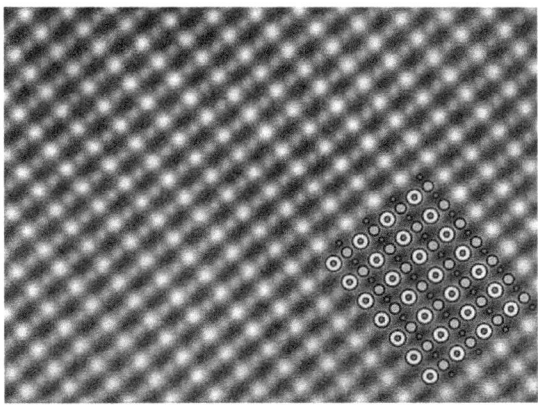

Atom resolution STEM image by Magnunor (Own work) CC BY-SA 4.0.

The illusion of the electron is a similar type of illusion. When an object moves close to the speed of light, time stops for that object. Quantum physics allows things without time to be everywhere they could typically be at the same time. So, why is the electron said to be everywhere in the electron cloud simultaneously?

5) What do you see between the proton and electron?

6) Quarks orbit inside protons, making them also appear solid. Is there anything solid inside the atom? Explain.

7) If atoms are mostly empty space, why do you think they look solid?

8) What would happen to the electron if time were to stop?

9) If that would happen to the electron, what would happen to the atom?

10) What would happen to all atoms in the area where time stops?

11) This answer is why we call this the space-time continuum. You cannot have space without time. When time disappears, so does space. Quantum mechanics shows that objects outside of space and time create objects in space and time. The Theory of Relativity $E=mc^2$ says matter can be converted into energy, and energy can be converted into matter. The way the atom is structured shows us this is not only possible but also real. The whole universe has and can fit inside of what structure?

　　a. We know all this is possible because the atom is mostly made up of what?

12) Why do we consider the atomic theory a theory? Why is it not a law or hypothesis?

WMAP

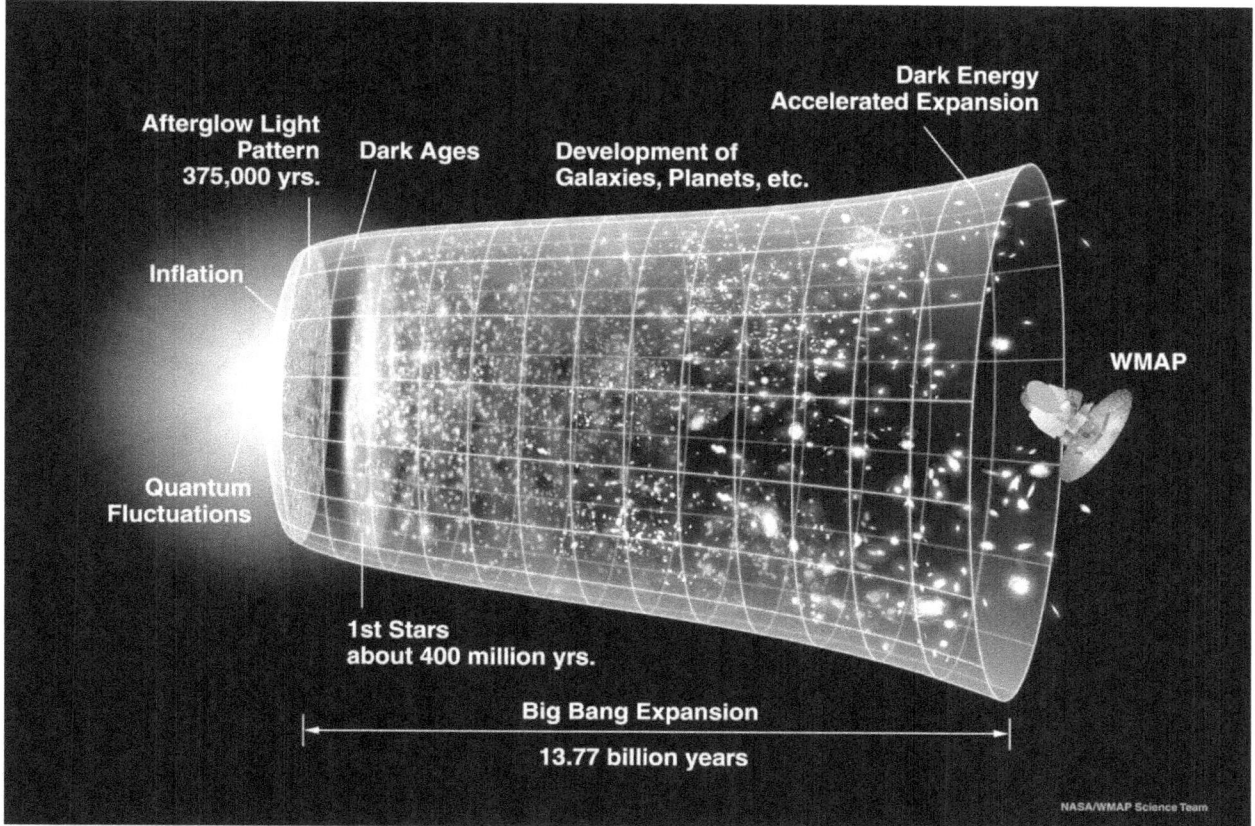

Directions:

Use NASA's image created by the Wilkinson Microwave Anisotropy Probe (WMAP) to answer the questions below.

1) How old is the universe?

2) What was the farthest this probe could look back(hint: Afterglow Light Pattern)?

3) How old was the universe when the first stars formed?

4) Describe how galaxies seem to change over time.

5) When does it look like the universe expansion started to accelerate?

6) What do physicists think is responsible for accelerated expansion?

7) What is accelerated expansion of the universe?

8) What do you think this means for the future of our universe? Discuss with class.

9) Could there be a time in the future that other galaxies will not be visible to us from Earth (even with the most powerful telescopes)? Explain.

Virtual Investigations that go with the Development of the Universe

ExploreLearning.com:

   Star Spectra

   H-R Diagram

   Orbital Motion – Kepler's Laws

   Nuclear Reactions

   Big Bang Theory – Hubble's Law

   Solar System

   Solar System Explorer

   Comparing Earth and Venus

Phet.colorado.edu:

   Blackbody Spectrum

   Gas Properties

   Gas Intro

   Gravity and Orbits

   Gravity Force Lab

   Gravity Force Lab: Basics

   Molecules and Light

Physicsclassroom.com:

   Interactives:

     Circular and Satellite Motion

     Orbital Motion

     The Value of g

Gravitation

Concept Builders:

Circular and Satellite Motion

Circular Logic

Case Studies – Circular Motion

Force and Free – Body Diagrams in Circular Motion

Universal Gravitation

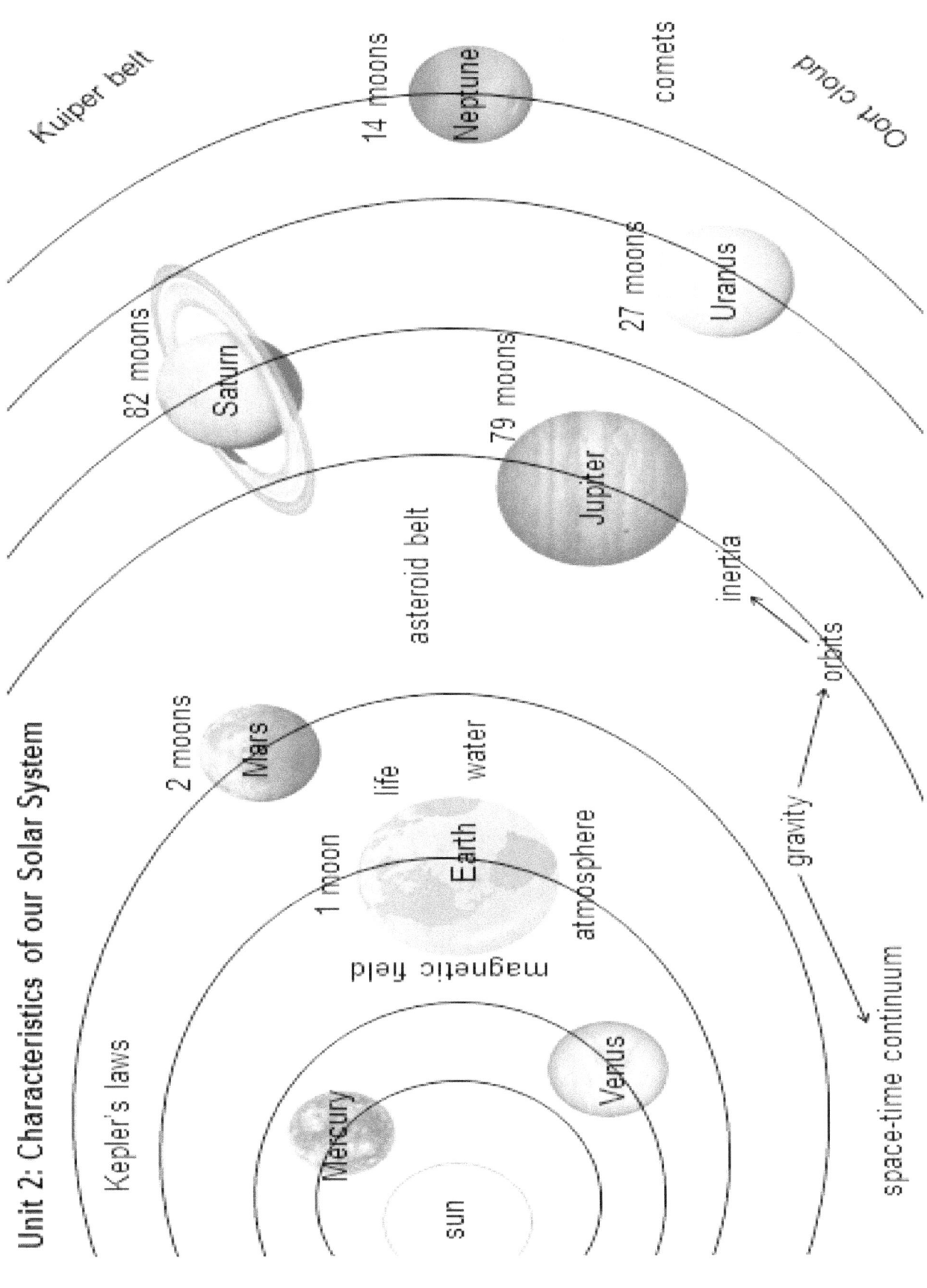

Simulating the Orbit of a Planet and Sun

Directions and Questions:

You will need a small **light ball** tied to a **string** and a **heavier ball** tied to a string. Your head will simulate the sun, and the ball will simulate a planet. **Looking at the materials and lab we will be using, what are the safety precautions we should take to protect ourselves and materials during the investigation?**

1) Hold the end of the string up near your head and swing the ball around your head. Notice your head moves back and forth, wobbling with the ball's motion; this is similar to how a sun wobbles to the orbit of a planet. Why do you think the sun wobbles?

2) Make sure you orbit the small ball, then the heavier ball around your head. Which one caused your head to wobble more?

3) What do you think causes the size of the wobble of a sun?

4) When looking at a star, what do you think a physicist could tell from the wobble of a star?

Bending of Space-time

Directions and Questions:

You will need the largest **Embroidery Hoop/Cross Stich Hoop** you can find. One for each group in your class. You will need to stretch some **elastic fabric** across and fix it into the hoops; this will represent the fabric of space-time. You will also need some **different weighted balls** and a **marble** to represent objects in the universe. **Looking at the materials and lab we will be using, what are the safety precautions we should take to protect ourselves and materials during the investigation?**

1) Have one student hold the hoop parallel to the ground. Have another student place the lightest ball inside the hoop in the fabric. What do you see happen to the fabric?

2) Now take the first ball off and place a heavier ball on the fabric. How was this different from the first ball?

3) How does this model show how objects bend the space-time continuum?

4) Roll a marble into your model with a heavy ball in the middle. What does it do to the larger object?

5) How does this show how gravity works in our universe?

6) Does this show gravity is a push or a pull? Explain why. Hint think of the ball in the center as like the Earth.

7) How would the fabric bend if you had an infinite mass in the center?

8) What would you make?

9) How is this model we used not accurate?

Kepler's Laws

Directions:

Use the **internet** and your **textbook** to describe Kepler's Laws and examples of each.

1) Kepler's First Law:

   a. Examples:

2) Kepler's Second Law:

   a. Examples:

3) Kepler's Third Law:

   a. Examples:

4) How can the orbits of objects change?

The Push of Gravity

1) Use the **internet** to find the masses of the **Sun** and **Mars** and the average distance between them. Use Newton's Universal Law of Gravitation $F = Gm_1 \times m_2/r^2$ ($G=6.674 \times 10^{-11}$) to help you calculate the gravitational force between them. Make sure you show your work in the formulas you used.

2) Use Newton's Universal Law of Gravitation to calculate the gravitational force between the **Sun** and **Jupiter**.

3) Which planet has a bigger effect on the Sun? Give evidence and explain why.

4) What are the variables that affect the force of gravity between objects?

5) Calculate the force of gravity between your body and the Earth.

6) How much would the force of gravity on your body change if you were on the surface of Mars?

7) If you were on the biggest moon of Mars, which object would have a bigger gravitational force on you, Mars or that moon? Give evidence and explain why.

8) Why does the title of this investigation say gravity has a push instead of a pull?

Observing Forces in Orbits (a)

Directions and Questions:

Tie a string to a tennis ball, and get a penny and a wire coat hanger. Looking at the materials and lab we will be using, what are the safety precautions we should take to protect ourselves and materials during the investigation?

1) Stand away from others, tightly hold one end of the string and spin the tennis ball around you above your head. Which direction does the ball want to go?

2) How can you tell?

3) How does this investigation show how and why the planets orbit the Sun and the moons orbit the planets?

4) Which object represents gravity?

5) Which object represents the sun?

6) Which object represents Mars going around the sun?

7) Which object represents a moon going around Jupiter?

8) What do you think causes the bubble of high tide between the Earth and the moon?

9) How does a planet's inertia affect its path of motion relative to the Sun?

10) Take the wire coat hanger and bend the triangle's bottom side to make a square so you can spin the hanger around your finger there. You may have to file down the end of the hook so you can balance a penny on it. Once the penny is balanced (works best tails side down), gently spin the hanger around your finger. The penny should stay balanced on the hanger if you do it right. Explain why this happens.

11) Which object is being orbited?

12) Which object is doing the orbiting?

13) Which object represents the force of gravity?

14) How does this model show how gravity works on orbiting objects?

15) How does this model not accurately show how gravity works on objects in orbit?

Solar System NASA Web Quest

Directions:

Use the Internet to go to the URL address www.science.nasa.gov/solar-system and search this site to answer the following questions about our solar system.

1) What are the ten things to know about the solar system?
    a.

    b.

    c.

    d.

    e.

    f.

    g.

    h.

    i.

    j.

2) What currently defines a planet?

3) How has the definition of a planet changed over time?

4) What are the planets of the solar system?

5) What are the dwarf planets that we know of?

6) What defines a dwarf planet?

7) How many moons does each of the planets have?

   a. Mercury

   b. Venus

   c. Earth

   d. Mars

   e. Jupiter

   f. Saturn

   g. Uranus

   h. Neptune

8) How do exoplanets compare with the planets of our solar system?

a. How do we detect exoplanets?

9) What are Asteroids?

   a. How many asteroids do we know of?

   b. What are the characteristics of asteroids?

   c. What are the different classifications of asteroids?

   d. What are asteroids composed of?

   e. Why did we send a probe to study Bennu?

       i. What is special about September 2135?

10) What is Didymos, and why is it significant?

11) What are comets?

   a. How many do we know of?

   b. What are their characteristics?

   c. Where do they come from?

   d. What are they composed of?

12) What are Meteors and Meteorites?

   a. What is the difference between them?

   b. Where do they come from?

   c. How many fall to Earth each day?

   d. When is the next meteor shower?

13) How does the Earth help scientists practice looking for stuff on other planets?

   a. How does it help study the geology of other planets?

   b. How does it help study volcanism in other worlds?

   c. How does it help study erosion in other worlds?

   d. How does it help study geologic impacts on other worlds?

   e. How does it help study tectonic activities in other worlds?

   f. How does it help study astrobiology on other planets?

   g. How does it help prepare for missions on other worlds?

14) What is the Kuiper Belt?

15) What are the ten things we need to know about the Kuiper Belt?

    a.

    b.

    c.

    d.

    e.

    f.

    g.

    h.

    i.

    j.

16) That is the Oort Cloud?

17) What are the ten things we need to know about the Oort Cloud?

    a.

    b.

    c.

    d.

    e.

f.

g.

h.

i.

j.

What is Planet X?

18) Where is it?

19) How big is it?

20) Why do we think it is there?

Scale Model of the Solar System

Directions and Questions:

You will need some **Play-Doh**, **meter sticks**, and **computers with internet** or a **textbook**. **Looking at the materials and lab we will be using, what are the safety precautions we should take to protect ourselves and materials during the investigation?**

1) Assign each group in the class to find the diameter of the sun, each planet of our solar system, and its distance from the sun.
2) With the help of your teacher, develop a scale to build your solar system in your classroom.
3) Take the Play-Doh and build the sun and the planets to that scale.
4) Using the same scale, measure how far each planet is from the sun and place them at those distances from the sun. What do you notice about this model?

5) What is the solar system mostly made up of?

6) Could the Kuiper belt fit in your classroom with this model?

7) Estimate how far it would be from the sun in your model.

8) Could the Oort cloud fit in your classroom with this model?

9) Estimate how far it would be from the sun in your model.

Scale Model of a Planetary System

Directions and Questions:

You will need **Play-Doh**, a **meter stick**, and a **textbook** or **computer** with the **internet**. **Looking at the materials and lab we will be using, what are the safety precautions we should take to protect ourselves and materials during the investigation?**

1) Assign each group a planetary system from our solar system.
2) Have the students use the internet or their textbook to find the diameters of their planet and moons and the distances those moons are from their planet.
3) Use Play-Doh to build a scale model of the planet and its moons at the correct distance from their planet on their lab table.
4) What planet were you assigned?

5) What is the diameter of this planet?

6) How many moons does it have?

7) Are there any other features your planet has? If so, what is it? Include this in your model.

8) Compare your planet to the others in the class. What is similar about the planets in the solar system?

9) What is different about the planets in our solar system?

10) How is this model not accurate?

Map of our Solar System

Directions and Questions:

You will need some **colored pencils**, **butcher paper**, and your **textbook** or a **computer** with the **internet**. **Looking at the materials and lab we will be using, what are the safety precautions we should take to protect ourselves and materials during the investigation?**

1) Use the internet or your textbook to draw a map of our solar system. Include the Sun, all the planets, important planetoids, moons of planets, the asteroid belt, important comets, the Kuiper belt, and the Oort cloud; this does not have to be to scale.
2) Find out how far away the closest star is to our sun. Which star is it?

3) Where would this star fit on this map?

4) Find other things that are in our solar system that were not mentioned in number one and describe them below.

Evolution of our Solar System

Directions:

You will need the **internet** to research how our solar system evolved to what it is today and answer the questions below.

1) What did the sun form from?

2) What was the force that brought it together?

3) How did the planets form?

4) What do we think was the shape of the solar system at the beginning?

5) What evidence shows us this?

6) Why are the planets closest to the sun smaller and rocky?

7) Why are the planets farthest away gas giants?

How we Know the History of the Earth

Directions:

Use the **internet** and your **textbook** to answer the questions below.

1) What evidence is there that the Earth's cooling led to tectonic activity, resulting in continents and ocean basins?

2) How old are the oldest continental rocks?

3) How old are the oldest rocks on the ocean floor?

   a. Why are these rocks so much younger?

   b. Where are the newest rocks forming?

      i. How is this affecting the size of an ocean?

4) Which of those rocks mentioned in numbers 1 and 2 would be better to find the age of the Earth? Explain why.

a. Which of the rocks are used to discover the history of the Earth? Explain why.

5) How do plate tectonics and erosion make finding the Earth's early history hard?

6) What other objects can scientists study that have not changed much over billions of years to help provide information about Earth's formation and history?

   a. Why would each of these objects be able to give us information about Earth's formations and history?

7) What does all this evidence tell us is the age of the Earth and the Solar System?

8) How did the impact of water-bearing comets on the Earth affect its atmosphere and hydrosphere?

The Greenhouse Effect on Different Planets

Directions:

You will need two **plastic tubs** (shoebox size) **painted black** on the inside, a **Press'n Seal sealing wrap**, two **temperature probes** attached to an **interface** connected to a **computer** with **Logger Pro**, and a light source like an **incandescent lamp** or some other lamp that gives off heat. **Looking at the materials and lab we will be using, what are the safety precautions we should take to protect ourselves and materials during the investigation?**

1) Make sure your tubs are both painted black on the inside, and a hole is poked through the end of each tub, big enough for a temperature probe to fit in snugly. Insert the temperature probes into the holes of both tubs.
2) On the first tub, make sure the Press'n Seal is secured to the tub's opening and the temperature probe is plugged into channel 1 of the interface.
3) The other tub needs to remain open, and the temperature probe plugged into channel 2.
4) In Logger Pro, open the folder Earth Science with Vernier and file #24 Greenhouse Effect.
5) Make sure your light source is equal distance from both tubs. Press "Collect" in the Logger Pro and turn on your lamp.
6) Monitor the time; when **5 minutes have passed, turn off the light**. Data will continue to be collected.
7) At **10 minutes, turn the lamp back on**. Data collection will continue until 15 minutes. At 15 minutes, Data collection will stop.
8) Look at the data collected in the Logger Pro and fill in Data Table 1 for the initial temperature, the temperature at 5 minutes, the temperature at 10 minutes, and the temperature at 15 minutes.
9) Then subtract the temperatures between Probe 1 and Probe 2 to get the temperature differences at different times. Write this information on the right side of Data Table 1.

Data Table 1

	Probe 1 Greenhouse	Probe 2 Control	Temperature Difference
0 Minute Temperature (°C)			
5 Minute Temperature (°C)			
10 Minute Temperature (°C)			
15 Minute Temperature (°C)			

Questions:

1) When the Lamp was on, which tub heated faster?

2) Give a possible explanation for your answer to number 1.

3) When the lamp was off, which tub cooled faster?

4) Give a possible explanation for your answer for number 3.

5) Which planets in our solar system are like tub #1? Explain why.

6) Which planets in our solar system are like tub #2? Explain why.

7) What characteristics of an atmosphere allow ice planets and moons?

8) The habitability of the Earth is a result of a delicate balance of the greenhouse effect. How/why is this statement true?

9) What could be sources of error in this investigation?

10) Do you think the Greenhouse Effect is a hypothesis or a theory? Explain why.

Seeing Patterns in the Layers of the Atmosphere

Directions:

Table 1 contains the average temperature readings at various altitudes in the Earth's atmosphere. Plot this data in Graph 1. Connect adjacent points with a smooth curve. Be careful to plot the negative temperature numbers correctly. This profile provides a general picture of temperature at any given time; however, the actual temperature may deviate from the average values, particularly in the lower atmosphere.

Data Table 1

Altitude (km)	Temp (°C)	Altitude (km)	Temp(°C)
0	15	52	-2
5	-18	55	-7
10	-49	60	-17
15	-56	65	-33
20	-56	70	-54
25	-51	75	-65
30	-46	80	-79
35	-37	85	-86
40	-22	90	-86
45	-8	95	-81
48	2	100	-72

1) Label the different layers of the atmosphere and the separating boundaries between each layer (words are listed below).

2) Mark the general location of the ozone layer.

*Read the background information on the next page after the graph and use it to place the eight words below on your graph in the correct locations:

troposphere, tropopause, stratosphere, stratopause, mesosphere, mesopause, thermosphere, and the **ozone layer.**

Graph 1

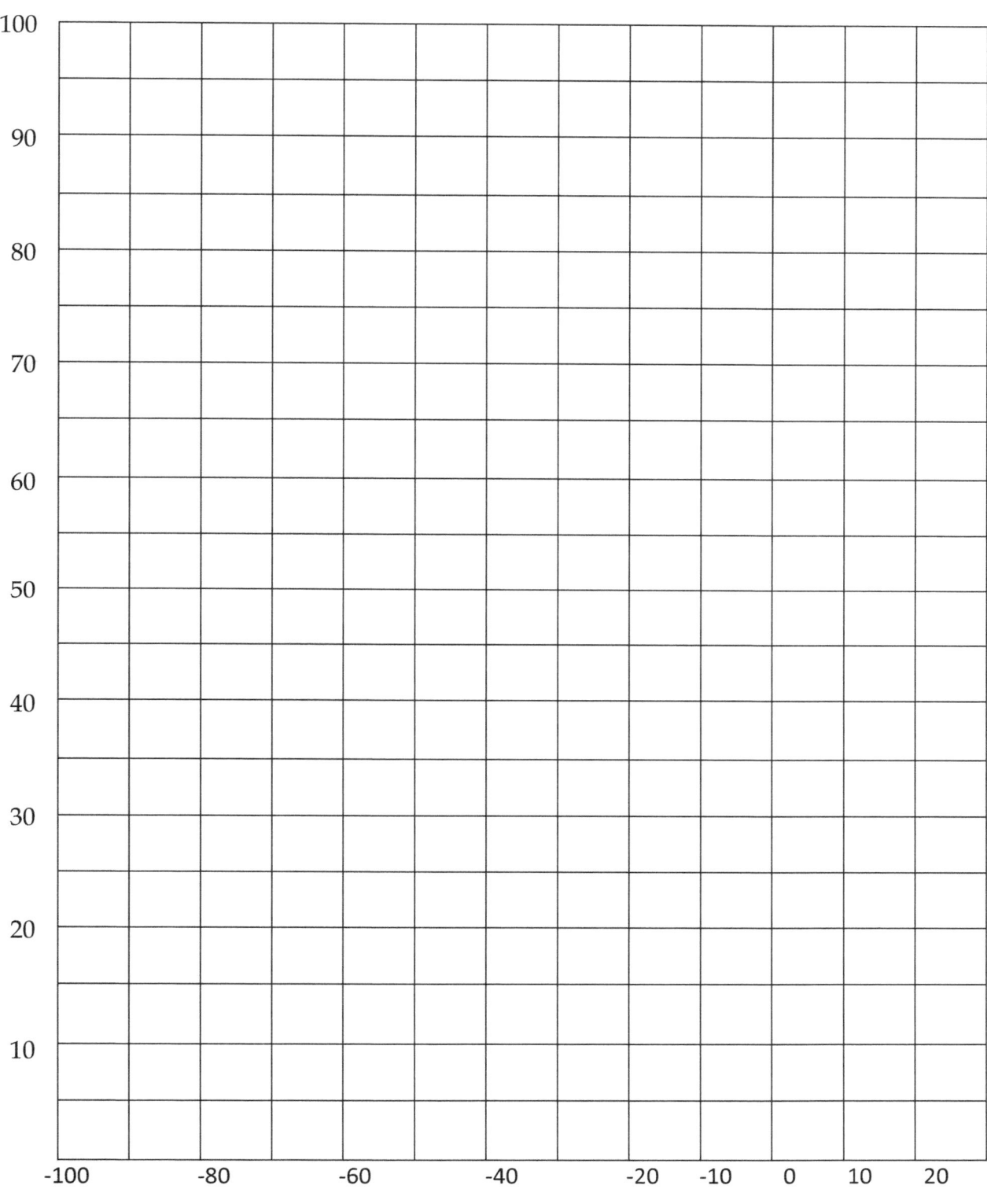

BACKGROUND:

The atmosphere can be divided into four layers based on temperature variations. The layer closest to the Earth's surface is called the **troposphere.** The **stratosphere** is above this layer, followed by the **mesosphere**, then the **thermosphere**. The upper boundaries between these layers are the **tropopause**, the **stratopause**, and the **mesopause**.

Temperature variations in the four layers are due to how solar energy is absorbed as it moves downward through the atmosphere. The Earth's surface is the primary absorber of solar energy. Some of this energy is reradiated by the Earth as heat, which warms the overlying **troposphere**. The **troposphere** is the lowest part of the atmosphere, the part where we live. It contains most of our weather, like clouds, rain, and snow. In this part of the atmosphere, the temperature gets colder as the distance above the Earth increases. The global average temperature in the **troposphere** rapidly decreases with altitude until the **tropopause**, the boundary between the **troposphere** and the **stratosphere**. It also contains about 75% of the air in the atmosphere and almost all water vapor, forming clouds and rain.

The temperature begins to increase with altitude in the **stratosphere**. This warming is caused by a form of oxygen called **ozone** (O_3) absorbing ultraviolet radiation from the sun. **Ozone** protects us from most of the sun's ultraviolet radiation, which can cause cancer, genetic mutations, and sunburn. Scientists are concerned that human activity contributes to a decrease in **stratospheric ozone**. Nitric oxide, which is in the exhaust of high-flying jets, and chlorofluorocarbons (CFCs), used as refrigerants, may contribute to **ozone** depletion.

At the **stratopause**, the temperature stops increasing with altitude. The overlying **mesosphere** does not absorb solar radiation, so the temperature decreases with altitude. Most of the meteors and rock fragments burn up in this layer.

At the **mesopause**, the temperature begins to increase with altitude, and this trend continues in the **thermosphere**. Here solar radiation first hits the Earth's atmosphere and heats it. Because the atmosphere is so thin, a thermometer cannot measure the temperature accurately, and special instruments are needed. This layer is relatively thin and is where space shuttles orbited and the space station orbits today. A small change in energy can cause a large change in temperature in this layer. The temperature in this layer can rise to 1,500 °C or higher.

Questions:

1) What is the basis for dividing the atmosphere into four layers?

2) Does the temperature increase or decrease with altitude in the:
   troposphere? _____ stratosphere? _____

   mesosphere? _____ thermosphere? _____

3) What is the approximate height and temperature of the:
   tropopause: _____ _____

   stratopause: _____ _____

   mesopause: _____ _____

4) What causes the temperature to increase with height through the stratosphere and decrease with height through the mesosphere?

5) What causes the temperature to decrease with height in the troposphere?

6) Describe the key characteristics of each atmospheric layer

Troposphere

Stratosphere

Mesosphere

Thermosphere

Composition of Planet Atmospheres

Directions:
Use the **internet** to list the top 5 chemicals and their % in the atmosphere for the different planets below(include how much the water range can be).

1) Mercury:

2) Venus:

3) Earth:

4) Mars:

5) Jupiter:

6) Saturn:

7) Uranus:

8) Neptune:

Atmospheres of Moons

Directions:

Use the **internet** to find the composition of the top three moons with the thickest atmospheres and answer the questions that follow. List the top 5 chemicals and their % in the atmosphere for the different moons you found (include how much the water range can be).

1) Titan

2) Triton

3) Io

4) Which planet does each of these moons orbit?

5) Why is it important to look at moons with atmospheres?

6) Why is water important for an atmosphere?

7) Which moons have oceans under the ice in our solar system?

   a. What could be in those oceans? Explain how it could be possible.

8) Explain what causes planets and moons to be able to hold on to their atmospheres?

9) What are the Van Allen belts, and how does the Earth's magnetic field form the Van Allen belts?

   a. What are auroras, and how do they form?

Models of Micro-molecules

Directions:

You will need a **molecular model kit** and a **Periodic Table**. Looking at the materials and lab we will be using, what are the safety precautions we should take to protect ourselves and materials during the investigation?

1) At the top of your periodic table, label it like this just below:

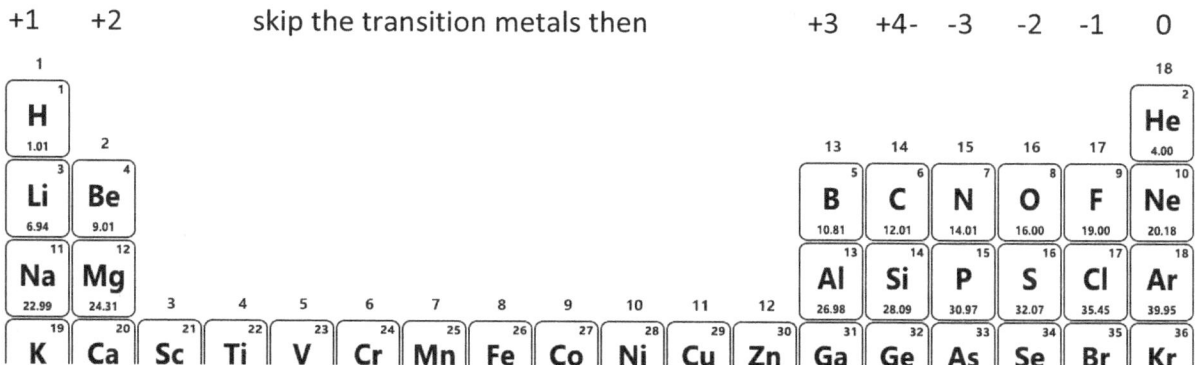

2) Different kits have different colors. In my kit, the:

   a. +1 (one-prong white) represents the Alkali Metals
   b. +2 (two-prong yellow) represents the Alkaline Earth Metals
   c. +3 (three-prong blue) represents the Boron Group
   d. +/- 4 (four prong black) represents the Carbon Group
   e. -3 (three-prong red) represents the Nitrogen Group
   f. -2 (two-prong blue) represents the Oxygen Group
   g. -1 (one-prong green) represents the Halogens
   h. The white tube is the bond

3) The different pieces in #2 represent the elements in those groups. Put the following molecules together:

H_2 (Make two): Take two hydrogen (white one-pronged) pieces connected by one white tube (single bond); this is the simplest molecule made and the most abundant in the universe.

O₂ (Make two) Take two oxygen (small blue two-pronged) pieces connected by two white tubes (double bond); this is **oxygen gas**; this makes up 18-20% of our atmosphere and is needed for aerobic respiration.

H₂O Take the two hydrogen molecules and the two oxygen molecules and rearrange them to make two **water molecules**. The oxygen (blue two-prongs) in the middle is connected with two white tubes to the two Hydrogen (white one-pronged) pieces (two single bonds); this is how rocket fuel burns to make water. If you do this in the opposite direction, this is how water separates during photosynthesis or electrolysis when electricity hits the water, and hydrogen and oxygen separate into their gases.

N₂ Take two nitrogen (red three-pronged) pieces and connect them to each other with three tubes (triple bond); this is nitrogen gas, the most abundant element in our atmosphere (78-80%).

NH₃ Take the one Nitrogen (red three-pronged) piece and connect it to three hydrogens (one-pronged white) pieces by three tubes (three single bonds); this is **ammonia**.

CO₂ Take one carbon (black four-pronged) piece and connect it to two oxygen (blue two-pronged) pieces with four white tubes (two double bonds); this is **carbon dioxide**.

CH₄ Take one carbon (black four-pronged) piece and connect it to four hydrogens (white one-prong pieces) with four white tubes (four single bonds); this is **methane gas**, the base for all fuels. **Methane** is a byproduct of decomposition away from oxygen. It is in all farts. Methane is also found in abundance away from Earth in the universe.

Questions:

1) What determines how atoms will combine with other atoms?

2) What do you notice about each of these molecules when building them?

3) How many atoms are in the largest molecule we made? This size is why we call them micro-molecules; they are very small.

Building a Model of a Water Molecule

Directions:

You will need a **balloon**, a **molecular model kit,** and a **Periodic Table**. Looking at the materials and lab we will be using, what are the safety precautions we should take to protect ourselves and materials during the investigation?

1) At the top of your periodic table, label it like this just below:

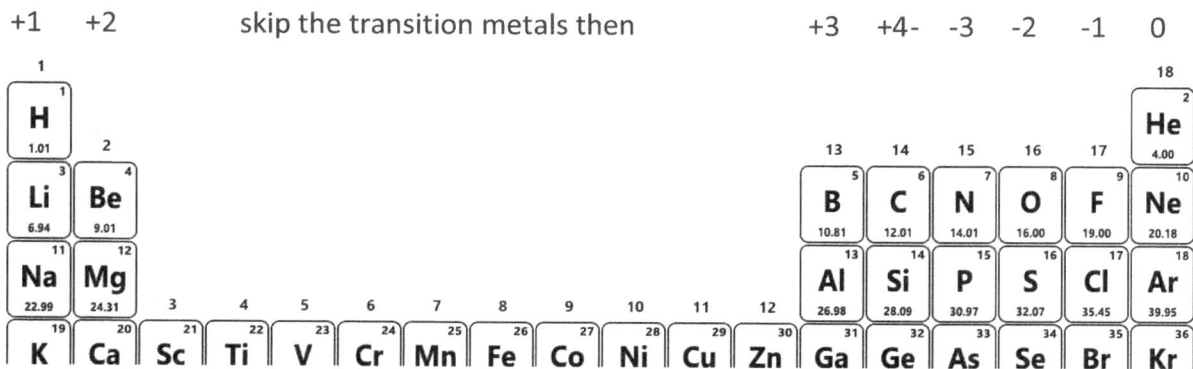

2) Different kits have different colors. In my kit, the:
   a. +1 (one-prong white) represents the Alkali Metals
   b. +2 (two-prong yellow) represents the Alkaline Earth Metals
   c. +3 (three-prong blue) represents the Boron Group
   d. +/- 4 (four-prong black) represents the Carbon Group
   e. -3 (three-prong red) represents the Nitrogen Group
   f. -2 (two-prong blue) represents the Oxygen Group
   g. -1 (one-prong green) represents the Halogens

3) Use the pieces to make two H_2O molecules. The hydrogen side of the molecule is slightly positive, and the oxygen side of the molecule is slightly negative making it polar like a magnet.

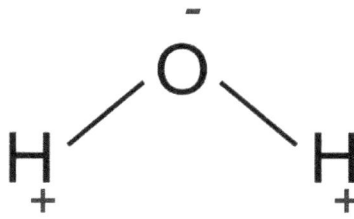

4) Because the water molecule has positive and negative ends, **ions** are attracted to the opposite charges on the water molecule. The positive ions are attracted to the oxygen side, and the negative ions are attracted to the hydrogen side. The same is true for other **polar molecules**; this is why ionic compounds and polar molecules like to dissolve in water. We call water the **universal solvent**.

5) Make a model of liquid water by taking your two water molecules and placing them next to each other where the oxygen of one is sitting between the two hydrogens of the other; this is how water molecules like to stick to each other. The positive ends are attracted to the negative ends; this is why water is **cohesive** (it sticks together).

6) When there are a bunch of them together, they have an equal pull on each other except for the ones on the surface; they are pulled slightly down because they have a slight charge above them. After all, there are no other molecules above them; this is why water has **surface tension**.

7) Since it is charged on both ends, it is also attracted to surfaces like a balloon with a static charge is attracted to a sweater or a wall. This attraction is why see water clinging to the sides of cold cans or glasses of ice tea. This phenomenon is called **adhesion**.
    a. You can model this by taking an inflated balloon, rubbing it on your hair to steal some electrons, and then sticking it to a shirt or wall.

8) You can make a model of ice (solid water) by flipping one of your water molecules and facing the oxygen ends toward each other. When water gets cold, the molecule's charge is not as strong, and the opposite ends are not attracted to each other anymore, so the oxygen atoms come together and share electrons with each other bonding them together. This orientation gives the molecule more space inside it and is why ice floats in liquid water.

Question:

1) Why do you think life depends so much on water?

2) How did these models show the characteristics of water?

3) How were the models inaccurate?

Models of Macromolecules

Directions:

You will need a **molecular model kit** and a **Periodic Table**. Looking at the materials and lab we will be using, what are the safety precautions we should take to protect ourselves and materials during the investigation?

1) At the top of your periodic table, label it like this just below:

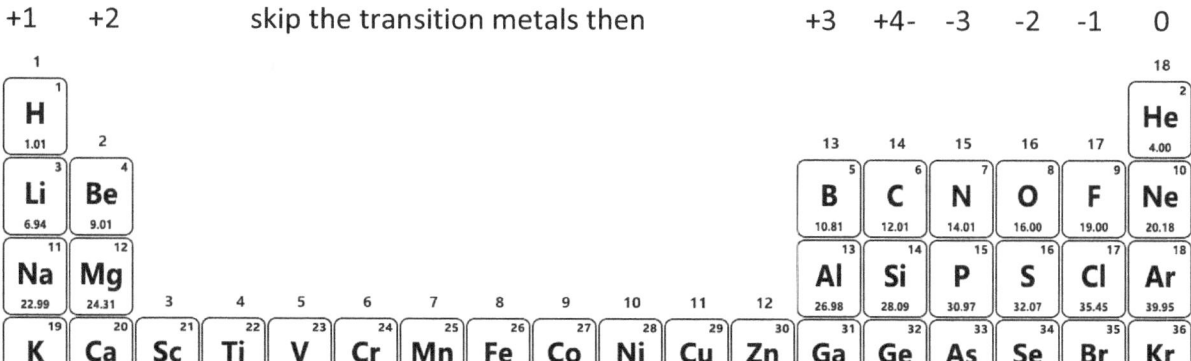

2) Different kits have different colors. In my kit, the:

   a. +1 (one-prong white) represents the Alkali Metals
   b. +2 (two-prong yellow) represents the Alkaline Earth Metals
   c. +3 (three-prong blue) represents the Boron Group
   d. +/- 4 (four prong black) represents the Carbon Group
   e. -3 (three-prong red) represents the Nitrogen Group
   f. -2 (two-prong blue) represents the Oxygen Group
   g. -1 (one-prong green) represents the Halogens
   h. The white tube is the bond

3) The different pieces in #2 represent the elements in those groups. Use pictures of large molecule monomers from your textbook or the internet to help you put them together. Then follow the directions to put monomers together to make polymers.

   a. **Carbohydrates**: the **monomer** is sugar like **glucose**. A **polymer** could be **starch**. When you make a chain of sugars by taking off a hydrogen atom from one sugar and OH from the other, this will produce water. You can then put the two sugars together to start a chain. This process is called **dehydration synthesis**. When you separate the starch back into sugars (like in digestion), water gets taken apart, and the H and OH get put back into each sugar molecule; this is called **hydrolysis**. Model this with your group. Sugar is used as fuel in living things.

b. <u>Lipids</u>: **Glycerol** and **fatty acids** are the **monomers**. They make membranes and store energy.

c. <u>Protein</u>: **Amino acids** are the **monomers** put together with **peptide bonds,** the building blocks of cell parts, enzymes, and hormones. An average usable protein is 300-400 amino acids long.

d. <u>DNA & RNA</u>: The **monomers** are **nucleotides** made of **sugar, phosphate**, and a **nitrogenous base**. These molecules can end up being 6 feet long in DNA. DNA is the genetic material for storing life's information to help build protein chains. Three types of RNA: **mRNA** (messenger RNA), **tRNA** (transfer RNA), and **rRNA** (ribosomes), all of which are needed to build proteins. If you are missing any of these, protein cannot be made, so; life cannot be made.

Questions:

1) From what you have observed in this lab, what are monomers?

2) What are polymers?

3) How many atoms are in the smallest molecule you made in this lab?

4) Can you count how big the largest would be?

5) If we were to build a cell with these model pieces, how big do you think it would be?

6) Is the chemical organization of life simple or complex? Explain.

How is Life Allowed on Earth?

Directions:

Use your **textbook** and **internet** as resources and what you have learned to explain the characteristics of the Earth that allow life to exist on it. Use this to answer the questions below:

1) What kind of sun/star do we need?

2) How far away does a planet need to be?

3) What effect does the rotation of a planet have on the ability for life?

4) How would a tilt of the axis affect the planet?

5) How does the moon help the Earth support life?

6) How does the atmosphere support life?

7) How does the magnetic field help the atmosphere on Earth?

   a. How did the Earth's magnetic field form?

8) What materials are needed for life to exist and thrive?

9) How would photosynthesis and aerobic respiration use these resources to help life thrive?

10) What can we learn about Venus and Mars to give us clues about the Earth?

11) What is the Goldilocks Zone, and how does that help life?

12) What things in the universe could destroy all life on Earth?

13) What protections are we getting from our solar system?

14) How complicated is the structure of life?

15) What does it take to make one cell the basic unit of life?

16) After researching this, how fragile is life on Earth, and how likely do you think we will find life on other planets?

17) Would it look like life on Earth? Explain why.

Virtual Investigations that go with Characteristics of our Solar System

ExploreLearning.com:

Solar System

Solar System Explorer

Gravity Pitch

Orbital Motion – Kepler's Laws

Uniform Circular Motion

Comparing Earth and Venus

Greenhouse Effect

Greenhouse Effect Metric

PhET.colorado.edu

Gravity and Orbits

Gravity Force Lab

Gravity Force Lab: Basics

Physicsclassroom.com

Physics Interactives

Circular and Satellite Motion

Uniform Circular Motion

Orbital Motion

Gravitation

The Value of g

The Value of g on Other Planets

Your Weight on Other Planets

Concept Builders

Circular and Satellite Motion

    Circular Logic

    Case Studies – Circular Motion

    Forces and Free–Body Diagrams in Circular Motion

    Universal Gravitation

    Gravitational Field Strength

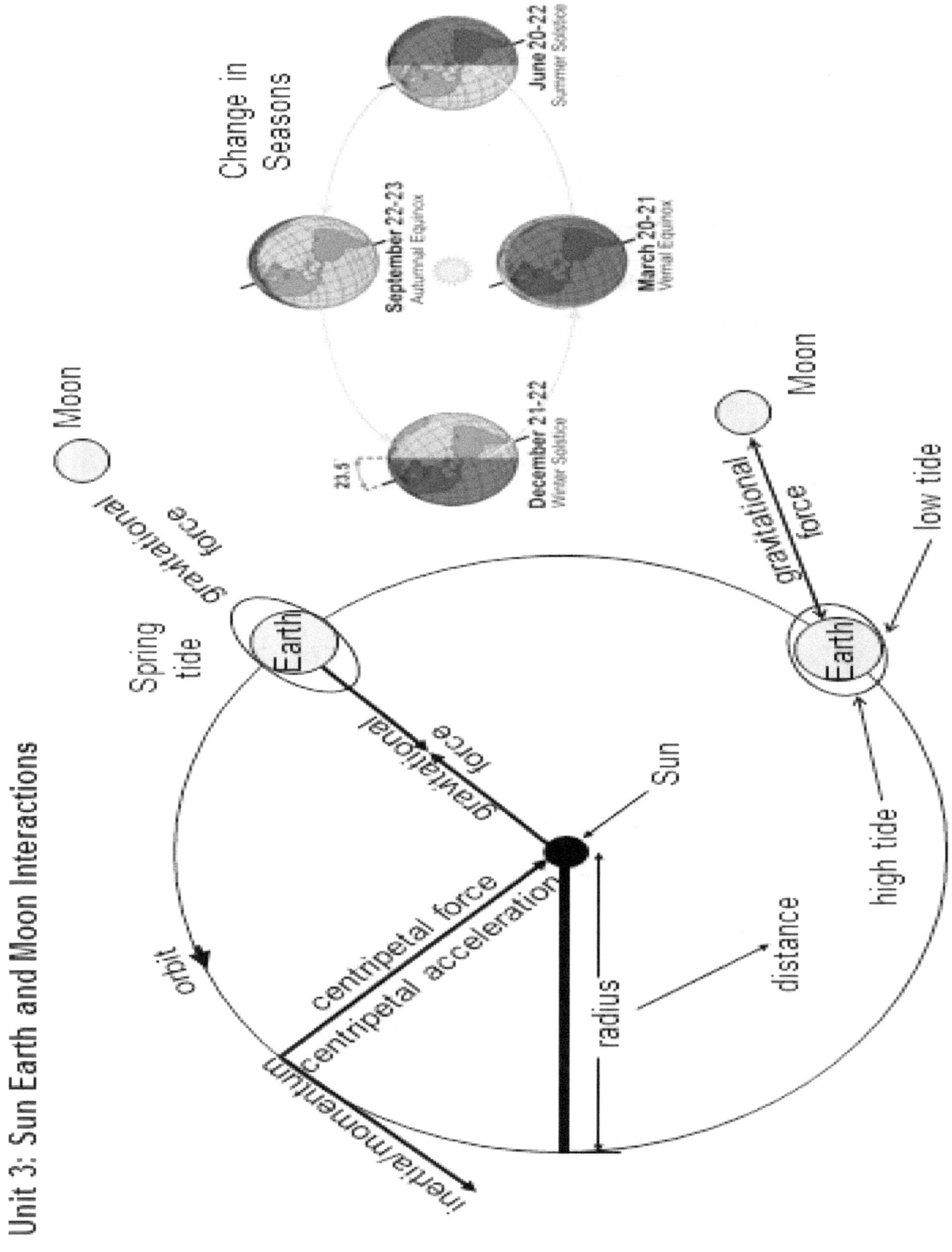

Seasons and the Tilt of the Earth

Directions:

You will need a **150-watt incandescent work lamp**, a **ring stand**, a **clamp** for the light source, a **globe**, **tape** that will hold the probe to your globe but not damage it, and a **temperature probe** attached to an **interface** connected to a **computer** with **Logger Pro**. **Looking at the materials and lab we will be using, what are the safety precautions we should take to protect ourselves and materials during the investigation?**

1) Set up your light on the ring stand so that it is pointing at the globe. Ideally, you want the bulb even with the middle of the globe. Place the bulb about 20 cm away from the globe.
2) Find your city on the globe and tape the temperature probe to the globe so that the probe's tip is over your city. Place the tape about 1 cm from the tip of the probe. Fold a piece of paper and wedge it under the back end of the temperature probe to keep it in contact with the globe's surface.
3) Position the globe so it is winter for the northern hemisphere. The North Pole will be tilting away from the light. Open the folder for Earth Science with Vernier and file #29 Seasons.
4) Click the "Collect" button, then immediately turn on the lamp. It will collect data for 5 minutes. When the data collection stops, turn off the lamp. Choose store latest run from the Experiment menu. (Do not touch the bulb; it will be very hot!)
5) Let the bulb, globe, and temperature probe cool to the initial temperature in your first data run.
6) Position the globe so it is summer for the northern hemisphere. The North Pole will be tilting towards the light. Click the "Collect" button and then turn on the lamp. It will collect data for 5 minutes. When the data collection stops, turn off the lamp.
7) Click the Statistics button, and then click OK to display statistics boxes for both runs. Record the minimum and maximum temperatures for the winter and summer runs and calculate the temperature change for both in Data Table 1.

Data Table 1

	Winter	Summer
Maximum Temperature (°C)		
Minimum Temperature (°C)		
Temperature Change (°C)		

Questions:

1) During winter in the northern hemisphere, is the light from the lamp able to hit the North Pole?

   a. How does this show there are 24 hours of the night during the winter?

2) Look at the South Pole. If the globe turns, will it ever go out of the light?

   a. How does this show there are 24 hours of the day during the summer?

3) How does this change when the Earth is positioned during the summer for the northern hemisphere?

4) Which season had a higher maximum temperature?

5) How does the temperature change for summer compared to winter?

6) Which season is the sunlight more direct?

7) What would happen to the temperature changes if the Earth were tilted more than 23.5°?

8) What causes the temperature and seasonal changes on Earth?

9) What other factors do you think could affect the weather in a region?

10) Discuss with your class other factors that can cause gradual climate changes. What are they?

    a. What are Milankovitch cycles, and how do they impact these gradual changes?

    b. What is the albedo effect, and how can this affect climate change?

The Pull of Gravity on Earth

1) Use the **internet** to find the masses of the **Earth** and its **moon** and the average distance from the moon to the Earth. Use Newton's Universal Law of Gravitation: **F = Gm$_1$ x m$_2$/r^2** (G=6.674 x 10^{-11}) with the help of your teacher to calculate the gravitational force between them. Make sure you show your work in the formulas you used.

2) Use Newton's Universal Law of Gravitation again to calculate the gravitational force between the **Sun** and the **Earth**.

3) Which object has a bigger effect on the tides of the Earth's oceans, the Sun or the moon? Give evidence and explain why.

4) Draw a force diagram showing the Earth, moon, and sun lined up in a row (showing a spring tide). What are the combined forces acting on the tides of the Earth?

5) Draw a force diagram showing the moon and sun make a right angle with the Earth (this is a neap tide). Subtracting the forces shows the effect the sun and moon have on the tides of the Earth.

6) Which kind of tide will pull more water, making the tides deeper?

7) Which kind of tide will not show as much change in tides during the day?

8) What are the variables that affect the force of gravity between objects?

Gravitational Interactions of the Sun and Moon on Earth

Directions:

You will need one large ball like a **soccer** or **basketball** for the whole class. You will need a **tennis ball**, a **marble** and a **piece of paper** for each group. **Looking at the materials and lab we will be using, what are the safety precautions we should take to protect ourselves and materials during the investigation?**

1) Have the large ball seen in the center of the room; this will represent the sun.

2) A tennis ball and marble will be at each table with each group. The tennis ball represents the Earth, and the marble represents the moon. Place a dot on the tennis ball to represent where someone is standing on the Earth.

3) Have the student arrange the Earth and moon showing a **spring tide**. The sun Earth and moon should be lined up in a straight line.

4) Have the students rearrange the Earth and moon, showing another **spring tide**.

5) Have the students arrange the Earth and moon showing a **neap tide**. The moon and the sun should be making a right angle with the Earth.

6) Have the students rearrange the Earth and moon, showing another **neap tide**.

7) There is always a bubble/swelling of water between the Earth and the moon and on the other side of the Earth away from the moon. Place a piece of paper under the Earth and draw this double bubble lined up with the earth and moon; this would show where the high and low tides are on the Earth.

8) Why do you think the double bubble follows the moon?

9) How would the sun affect the double bubble?

10) Slowly spin your Earth (the tennis ball) on its axis, allowing the ball's dot to spin around the ball. If someone were standing on this part of the planet, what would they observe about the water levels on Earth during one day?

11) Turn the dot on the ball to a **high tide**.

12) Turn the dot on the ball to a **low tide**.

13) How many times would someone experience a high tide in a day?

14) How many times would someone experience a low tide in a day?

15) How does this model show the interactions of the sun, Earth, and moon?

16) What is not accurate about this model?

17) Extra Credit: Which phases of the moon would show spring tides? Explain why.

18) Extra Credit: Which phases of the moon would show neap tides? Explain why.

Observing Forces in Orbits (b)

Directions and Questions:

Tie a **string** to a **tennis ball**, and get a **penny** and a **wire coat hanger. Looking at the materials and lab we will be using, what are the safety precautions we should take to protect ourselves and materials during the investigation?**

1) Stand away from others, tightly hold one end of the string and spin the tennis ball around you above your head. Which direction does the ball want to go?

2) How can you tell?

3) How does this investigation show how and why the Earth orbits the Sun and the Moon orbits the Earth?

4) Which object represents gravity?

5) Which object represents the sun?

6) Which object represents the Earth going around the sun?

7) Which object represents the moon going around the Earth?

8) What do you think causes the bubble of high tide between the Earth and the moon?

9) How does the moon's inertia affect the moon's path of motion relative to the Earth?

10) Take the wire coat hanger and bend the triangle's bottom side to make a square so you can spin the hanger around your finger there. You may have to file down the end of the hook so you can balance a penny on it. Once the penny is balanced (works best tails side down), gently spin the hanger around your finger. The penny should stay balanced on the hanger if you do it right. Explain why this happens.

11) Which object is being orbited?

12) Which object is doing the orbiting?

13) Which object represents the force of gravity?

14) How does this model show how gravity works on orbiting objects?

15) How is this model not accurately showing how gravity works on objects in orbit?

Orbit Under Glass

Directions:

You will need a **glass** or a **see-through plastic cup** and a **marble** on a tabletop. **Looking at the materials and lab we will be using, what are the safety precautions we should take to protect ourselves and materials during the investigation?**

1) Place the glass/plastic cup over the marble. Move the glass/cup in a circular motion causing the marble to spin around the inside of the glass/cup off the table's surface.
2) Discuss what you are seeing with the class and your teacher.

Questions:

1) Explain why the marble can stay off the surface of the table.

2) Which part of this lab setup represents an object like the Earth moving around the Sun?

3) Which part of the lab setup represents the force of gravity on these orbiting objects?

4) Where in the lab setup would be the object that is being orbited?

5) What happens when you stop the movement of your glass/cup?

6) How would this lack of inertia affect an orbiting object? Explain why?

Moon Phases

Directions and Questions:

You will need a **ping pong ball** with one half of it painted black for each student or group and a **large ball** sitting in the center of the room to represent the sun. **Looking at the materials and lab we will be using, what are the safety precautions we should take to protect ourselves and materials during the investigation?**

1) The person will be the Earth, and the ping pong ball will be the moon. Have the white part of the moon face the sun (the large ball in the middle of the room) and the moon's dark side face away from the sun.
2) The diagram below shows how the moon appears with the Earth at this angle at the eight different points shown.

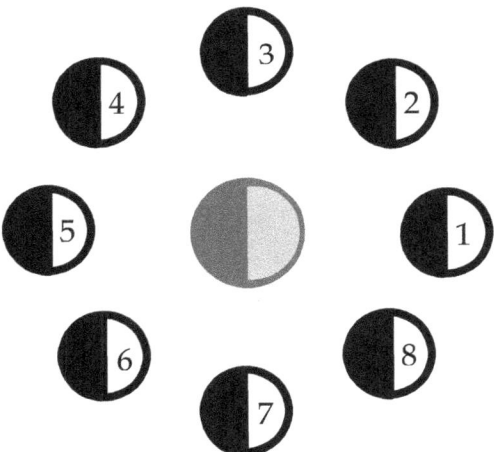

3) Arrange your ping pong ball (moon) and yourself (Earth) with the sun (large ball in the center of the room) in each of these positions above and shade how each phase of the moon would appear to you from Earth.

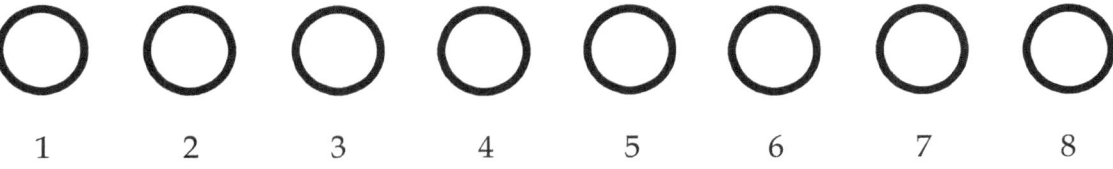

4) Which phase # is a new moon?

5) Which phase # is a full moon?

6) Which phase # is the first quarter?

7) Which phase # is the third quarter?

8) Which phase #s would have a neap tide? Explain why.

9) Which phase #s would have a spring tide? Explain why.

10) Explain why we see different phases of the moon.

11) How is this model accurate in showing us the moon phases?

12) How is the model not accurate in showing us the moon phases?

Origins of the Moon

Directions and Questions:

Use the **internet** to research the moon's origin that orbits the Earth and answer the following questions.

1) What are the four leading hypotheses of how the moon formed? Explain each.

   a.

   b.

   c.

   d.

2) Which of these hypotheses do scientists think is what happened?

   a. What is the evidence?

   b. How would this set the Earth up to support life?

Solar Eclipses

Directions and Questions:

You will need a tape measure, a **golfball** to simulate the moon, and a **small youth basketball** or **soccer ball** to simulate the Earth on a sunny day. You could also use the scale model of your planetary Earth system from a previous lab. **Looking at the materials and lab we will be using, what are the safety precautions we should take to protect ourselves and materials during the investigation?**

1) Find the scale distance the golfball (moon) should be away from the small youth basketball (Earth) if the moon and Earth were these sizes.
2) Move the golfball (moon) that scaled distance away from the basketball (Earth) between the sun and the basketball.
3) Move the golfball (moon), so you cast a shadow on the basketball (Earth). How much of the Earth is in the shadow of the moon?

4) When a solar eclipse happens, does everyone on Earth get to experience it? Explain.

5) Why is it so rare for someone to experience a solar eclipse?

6) What phase of the moon can a solar eclipse happen? Explain why.

Solar Eclipse on Earth Perspective

Directions and Questions:

You will need **eclipse glasses** or a **welding mask** to look at the sun. Never look at the sun without protective eyewear that is made for this. The moon is a perfect distance from the Earth and a perfect size to just cover the sun but not the sun's corona to allow us to see it with a telescope. We will place a pen or pencil over the sun to simulate a partial eclipse. **Looking at the materials and lab we will be using, what are the safety precautions we should take to protect ourselves and materials during the investigation?**

1) Go outside on a sunny day and where your eclipse glasses. Look at the sun; what shape does it have?

2) Place the end of your pen/pencil in front of the sun to create a shadow over part of the sun. What are the different ways your pen/pencil can create a partial eclipse?

3) What shapes does the sun show with partial eclipses?

Lunar Eclipses

Directions:

You will need a tape measure, a **golfball** to simulate the moon, and a **small youth basketball** or **soccer ball** to simulate the Earth on a sunny day. You could also use the scale model of your planetary Earth system from a previous lab. **Looking at the materials and lab we will be using, what are the safety precautions we should take to protect ourselves and materials during the investigation?**

1) Find the scale distance the golfball (moon) should be away from the small youth basketball (Earth) if the moon and Earth were these sizes.
2) Move the small youth basketball (Earth) that scaled distance away from the golfball (moon) between the sun and the golfball.
3) Move the small youth basketball (Earth), so you cast a shadow on the golfball (moon). How much of the moon is in the shadow of the Earth?

4) When a lunar eclipse happens, does everyone on Earth get to experience it? Explain.

5) What phase of the moon can a lunar eclipse happen? Explain why.

6) Which happens more, a solar eclipse or a lunar eclipse? Explain why.

Virtual Investigations that go with Sun Earth and Moon Interactions

ExploreLearning.com:

Seasons: Why do we have them?

Seasons in 3D

Summer and Winter

Seasons: Earth, Moon, Sun

Ocean Tides

Tides

Phases of the Moon

Moonrise, Moonset, and Phases

Eclipse

3D Eclipse

2D Eclipse

Penumbra Effect

Comparing Earth to Venus

Solar System

Gravity Pitch

Orbital Motion – Kepler's Laws

PhET.colorado.edu:

Gravity and Orbits

Gravity Force Lab

Gravity Force Lab: Basics

Physicsclassroom.com

Physics Interactive

Circular and Satellite Motion

   Uniform Circular Motion

   Orbital Motion

   Gravitation

Concept Builders

Circular and Satellite Motion

   Circular Logic

   Case Studies – Circular Motion

   Forces and Free-Body Diagrams in Circular Motion

   Universal Gravitation

   Gravitational Strength

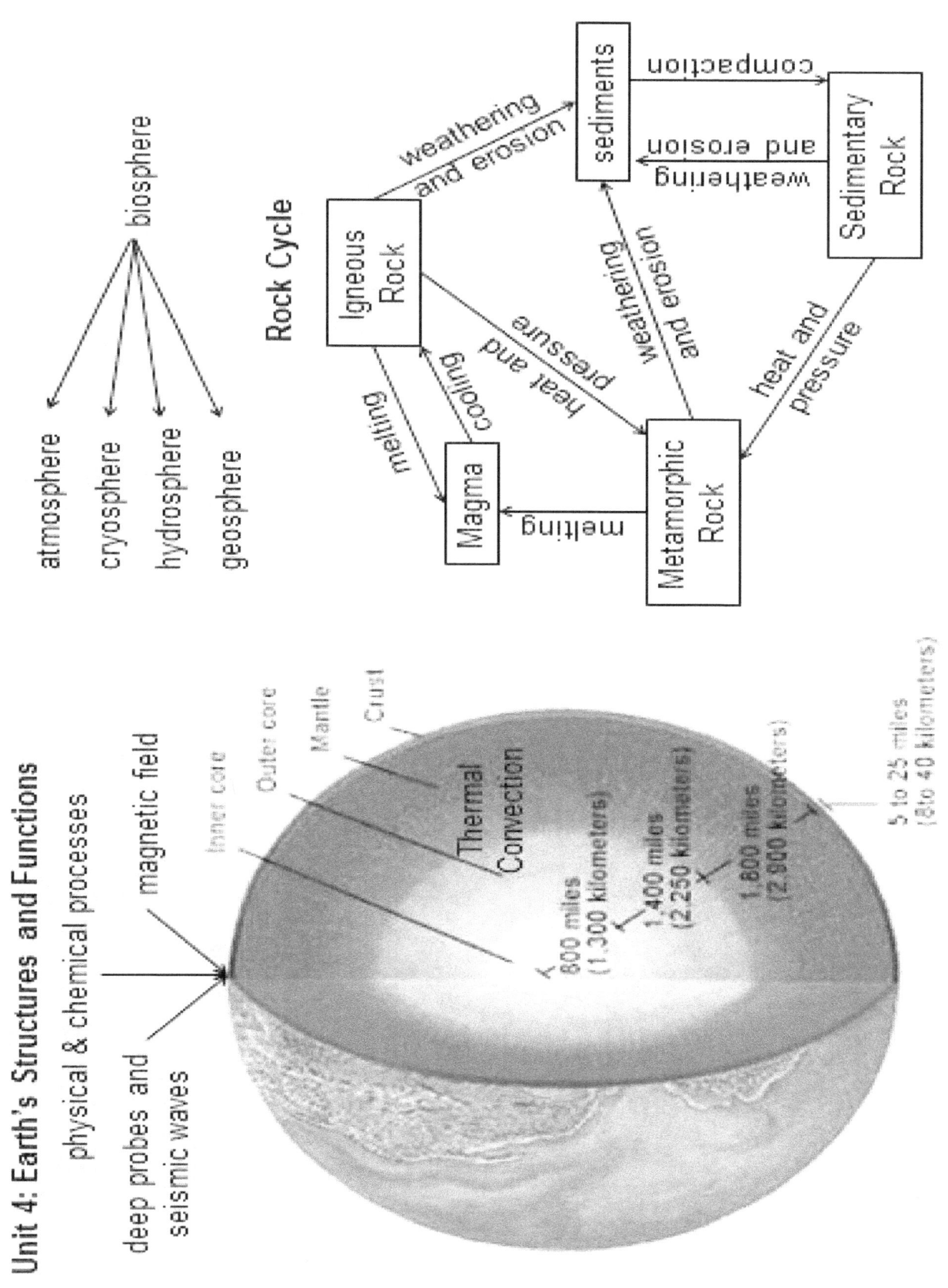

Unit 4: Earth's Structures and Functions

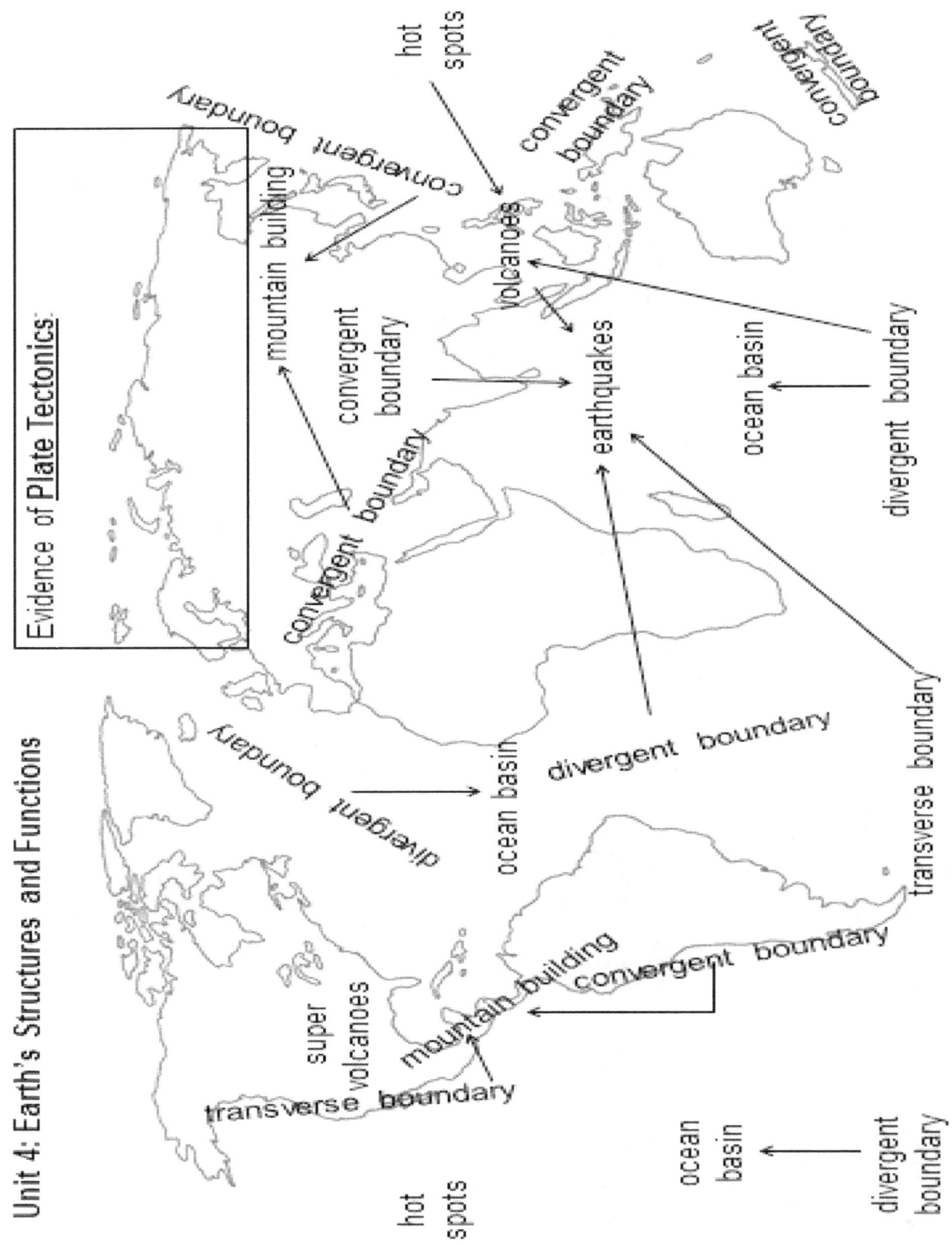

Observing the Different -spheres

Directions and Questions:

You will need to go outside to answer the questions below about the different parts of the Earth. **Looking at the materials and lab we will be using, what are the safety precautions we should take to protect ourselves and materials during the investigation?**

1) The **biosphere** is anywhere on Earth where life exists. Where do you see the **biosphere** outside?

2) Where else is the **biosphere** that you may not physically see but is near you?

3) Discuss with your class and write down the components and characteristics of the **biosphere**?

4) The **hydrosphere** is anywhere where there is liquid water. Where do you see the **hydrosphere** outside?

5) Where else is the **hydrosphere** that you may not physically see but is near you?

6) Discuss with your class and write down the **hydrosphere's** components, characteristics, and actions.

7) The **atmosphere** is where gases surround our planet. Where do you see the **atmosphere** outside?

8) Where else is the **atmosphere** that you may not physically see but is near you?

9) Discuss with your class and write down the atmosphere's components, characteristics, and actions.

10) The **geosphere** is all the rocks that make up the Earth (from mountains to the smallest grain of sand). Where do you see the **geosphere** outside?

11) Where else is the **geosphere** that you may not physically see?

12) Discuss with your class and write down the geosphere's components, characteristics, and actions.

13) How can the geosphere get worn away by the hydrosphere and atmosphere?

14) Which of the –spheres overlap, and how do they overlap?

15) Which of these –spheres do not overlap?

16) Depending on the season and where you live, you might see the **cryosphere** (the frozen parts of the planet where there is ice). How is the **cryosphere** similar to the other –spheres?

17) How is the **cryosphere** different from the other –spheres?

18) **Weathering** is the process of changing and breaking apart rocks. **Erosion** is when the pieces of the weathered rocks are moved away. **Deposition** is the process of dropping or settling sediment from weathering. How are weathering, erosion, and deposition involved in the interactions of all these spheres?

19) Where do you see evidence of weathering, erosion, and deposition taking place?

Modeling the Layers of the Earth

Directions and Questions:

You will need a **steel ball**, **Play-Doh** of three colors (red, orange, and brown), a **plastic knife**, a **scale**, and a **computer** with **internet. Looking at the materials and lab we will be using, what are the safety precautions we should take to protect ourselves and materials during the investigation?**

1) Take the steel ball, which is the solid iron inner core, cover it with 100 g of red Play-Doh (the liquid iron outer core), and make it into a ball.
2) Now take about 150 g of the orange Play-Doh and cover the red Play-Doh making a larger ball. You just added the mantel.
3) Now make a thin layer of brown Play-Doh and cover the orange layer. You just added the crust.
4) Take the plastic knife and cut your ball in half like an avocado. Look at the side with the steel ball; this is a model of the Earth's layers.
5) Use the internet to describe each layer of the Earth and fill it in below.

    a. Inner Core:

    b. Outer Core:

    c. Mantel:

    d. Crust:

6) What evidence did scientists use to develop our knowledge of the Earth's 3-D structure?

7) What are some radiometric dating methods used to calculate the ages of igneous rocks on Earth?

   a. How precise and reliable are these methods?

8) Describe how impact accretion, gravitational compression, radioactive decay, and cooling differentiate proto-Earth into layers.

9) Describe how heat and rock composition affect density within Earth's interior.

Soft Boiled Egg Earth

Directions and Questions:

You will need a **soft-boiled egg**, a **paper towel**, an **avocado** already cut in half, and a **knife**. **Looking at the materials and lab we will be using, what are the safety precautions we should take to protect ourselves and materials during the investigation?**

1) Take your soft-boiled egg and gently crack the shell repeatedly (like you are about to peel it, but do not peel it) so there are sections of the shell separated from other sections with cracks. How is the cracked egg shell like the crust of the Earth?

2) Take a knife or have your teacher take the knife and cut your egg in half on a paper towel. What part of the egg represents the solid inner core? Explain why.

3) The liquid part of the yolk (because it is soft-boiled) represents the outer liquid core. How is this part of the model similar to the earth?

4) Which part of the egg is the mantel? Explain why.

5) How is this a good model for showing the layers of the Earth?

6) How is this model inaccurate?

7) How would an avocado be a good model for the layers of the Earth (see the teacher's avocado)?

8) What parts of the avocado would be the...

   a. Core:

   b. Mantel:

   c. Crust:

9) How would an avocado be inaccurate?

Modeling the Rock Cycle

Directions and Questions:

You need **3-4 Starbursts** of different colors, **scissors**, a **Ziploc bag**, a **large hardcover book**, **aluminum foil**, **tongs**, a **hotplate**, and **safety goggles** for each student. You can use the Rock Cycle diagram on your concept map for this unit to help guide you through this investigation. **Looking at the materials and lab we will be using, what are the safety precautions we should take to protect ourselves and materials during the investigation?**

1) Unwrap your Starbursts. Each Starburst was melted and poured into a tiny mold to make its shape. What kind of rock would this represent?

2) Use your scissors to cut them up into tiny pieces. What does this process represent in the rock cycle?

3) Now take your tiny pieces and pile them up together. What part of the process is this in the rock cycle?

4) Squish them together with your hands using pressure, compacting the layers together. What kind of rock are we forming in this model?

5) What processes helped make this rock?

6) Now you will need to use heat and pressure, so take your model and place it in a Ziploc bag to trap the heat from your hands as you squish it more. Then place your large hardcover book on top and firmly press down, applying more pressure. You can take your model rock out and fold it upon itself, place it back in the plastic bag, squish it with your hands again, and then repress it with your book. Take the model rock out of the Ziploc bag. What type of rock did we just make in this model?

7) What was the process that made this rock?

8) Now make a bowl with the aluminum foil and place your model rock in it. Put your safety goggles on and place your bowl of aluminum foil and rock on a hotplate, then turn the hotplate on high and watch your model rock melt and boil. What part of the rock cycle are we modeling now?

9) After you see it boil, do not let it burn, turn off your hotplate and use some tongs to remove the bowl to cool. What type of rock did we just model?

10) What could you do to continue the rock cycle?

11) How could you change this model to skip steps to take different paths to other parts of the rock cycle?

12) If you have time, try them and see if they work. How does this show the rock cycle more accurately?

13) How was this a good model for showing the rock cycle?

14) How was this model inaccurate?

Nuclear Decay Half-life of Pennies (Fossils)

Directions:

You will need a **plastic tub** (about the size of a shoebox) with a **lid** and 100 **pennies**. **Looking at the materials and lab we will be using, what are the safety precautions we should take to protect ourselves and materials during the investigation?**

1) A radioactive element's half-life is how long it takes for half the atoms to change into another element as they go through radioactive decay. Since pennies have two sides, almost half will land on heads when flipped, and almost half will land on tails.
2) In your tub, place all 100 pennies heads up. These will represent your radioactive isotopes.
3) Place the lid on your box, shake it and count to 10.
4) Lift the lid and take out all the pennies that have landed tails up. Count them and fill in Data Table 1. Subtract the number of tails from 100 to show the number still heads up.
5) Now only the pennies that are heads up are still in the tub. Repeat the procedure in #s 3-4 six more times unless all your pennies turned up tails before that.
6) Graph your data from Data Table 1 on Graph 1.
7) Your teacher will give each group a number and put your data into Data Table 2 for your group. Get the data for the other groups and put them in Data Table 2
8) Average each half-life for all the groups by adding their data and dividing by the number of groups.
9) Graph the averages you have for the class in Data Table 2 on Graph 2.

Data Table 1

Shaking Time (s)	# of Heads	# of Tails taken out
10		
20		
30		
40		
50		
60		
70		

Data Table 2

Groups	# of Heads at 0 (s)	# of Heads at 10 (s)	# of Heads at 20 (s)	# of Heads at 30 (s)	# of Heads at 40 (s)	# of Heads at 50 (s)	# of Heads at 60 (s)	# of Heads at 70 (s)
1	100							
2	100							
3	100							
4	100							
5	100							
6	100							
7	100							
8	100							
9	100							
10	100							
11	100							
12	100							
13	100							
14	100							
15	100							
Average	100							

Graph 1

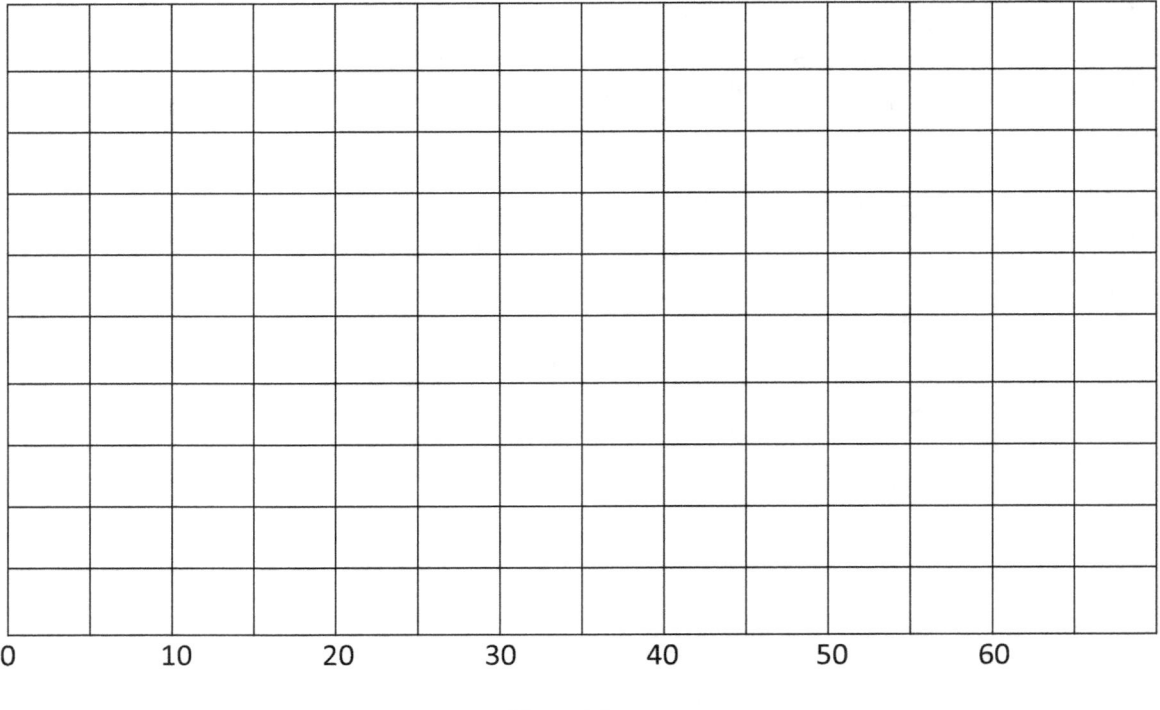

Time (s)

Graph 2

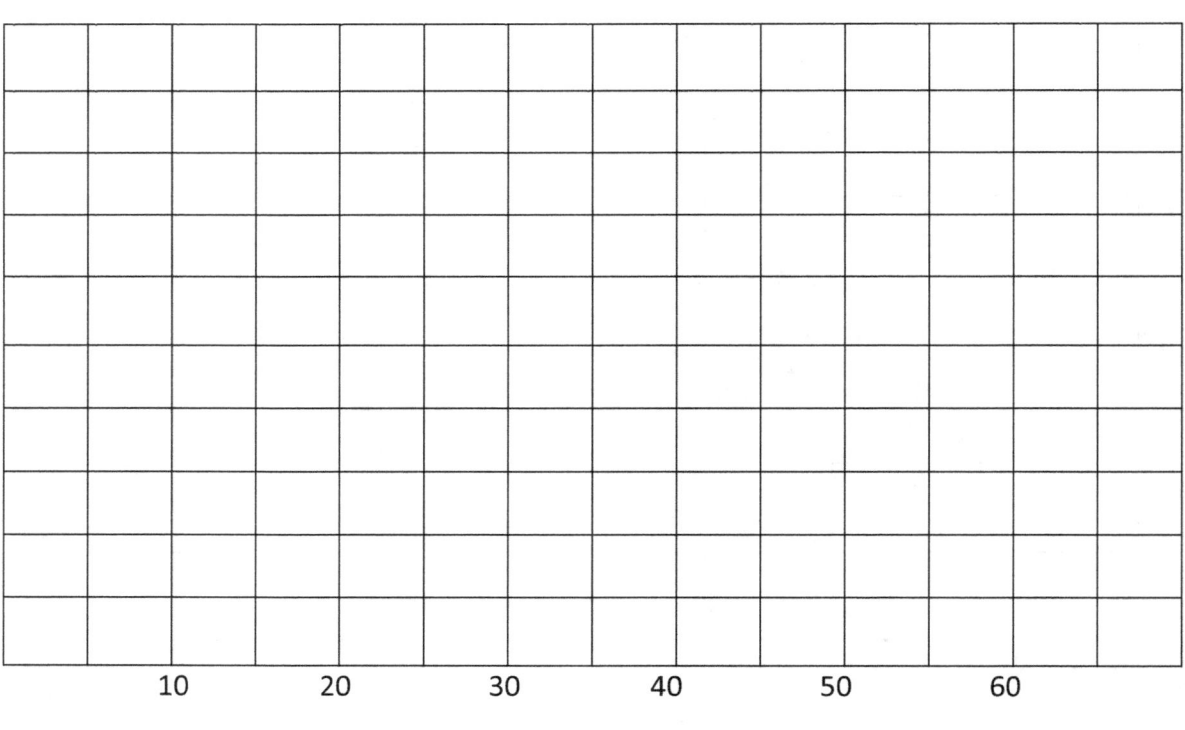

Time (s)

Questions:

1) What do the 10 seconds represent?

2) Which side of the coin was a stable isotope?

3) Which side of the coin was the unstable isotope?

4) What represented the radioactive decay?

5) How much of the atoms decay during each half-life?

6) How many half-lives until you should run out of pennies if all things go as expected?

7) If you started with 100 pennies and a half-life goes by every 10 seconds, how many pennies should there be after 40 seconds?

    a. How close was this to your group's results?

    b. How close was this to the class results?

    c. Why should the class results be closer to the expected?

8) How can the information in this lab be used to find the age of fossils?

Calculating Nuclear Half-life Decay (Fossils)

Directions and Questions:

Use **colored pencils** to color in the graph below as you follow the directions to simulate what half-life looks like.

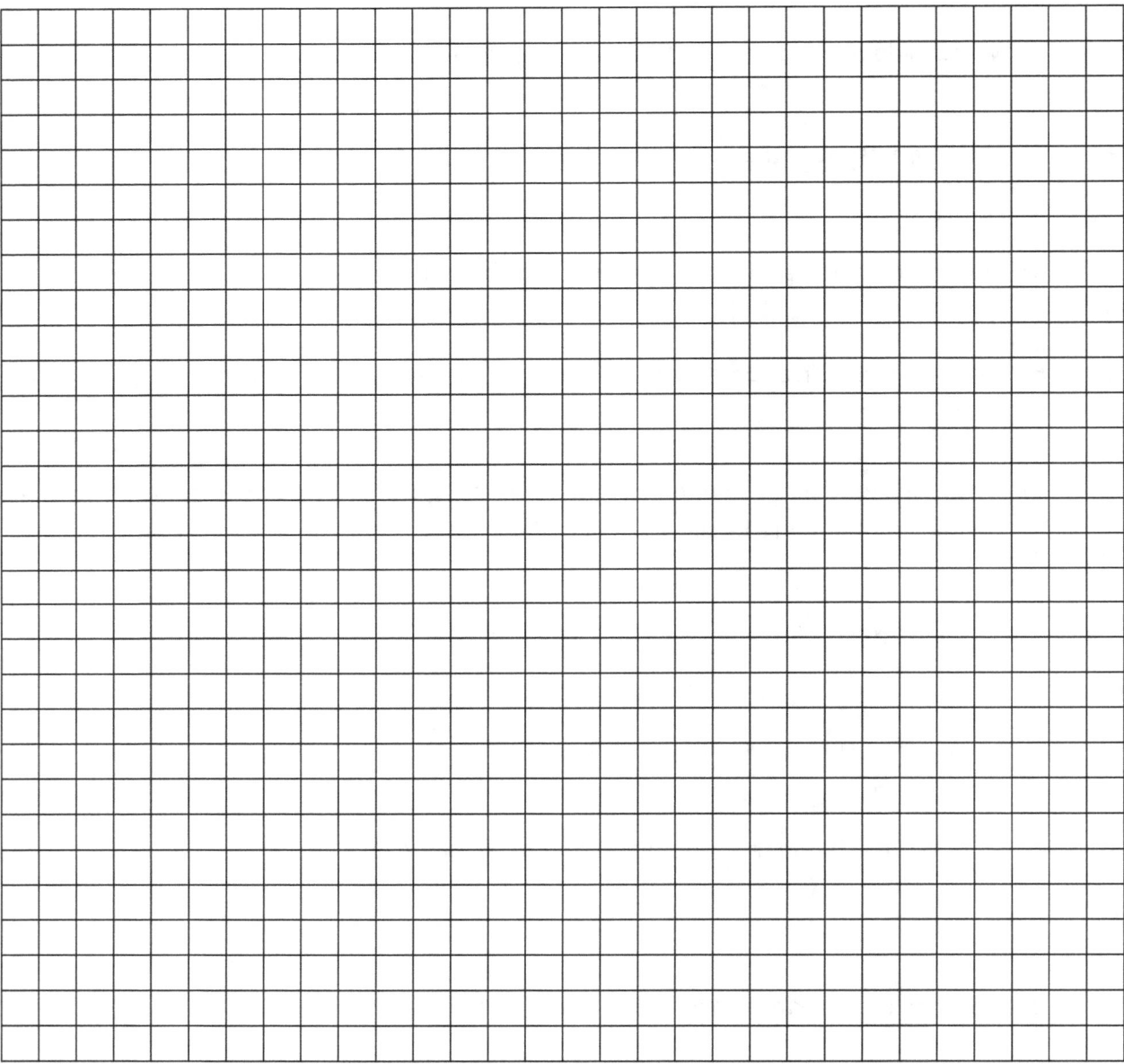

1) There are 900 squares above. Use a red colored pencil to shade in half of the squares. How many squares are left after one half-life?

2) Use a blue colored pencil to shade in half of the squares that are left; how many squares are left after two half-lives?

3) Use a green colored pencil and color in half of the squares now left. How many squares are left after three half-lives?

4) Use a yellow colored pencil and color in half of the squares now left. How many squares are left after four half-lives?

5) Use a purple colored pencil and color in half of the squares now left. How many squares are left after five half-lives?

6) Use a brown colored pencil and color in half of the squares now left. How many squares are left after six half-lives?

7) Suppose each square represented one atom of a substance that is decomposing. How many half-lives could go by until we should expect all the atoms to be gone? Explain why.

8) Carbon 14 is an isotope used to measure how long ago an organism died; It has a half-life of 5,730 years. It will work for about 50,000 years. How many half-lives is this?

9) For fossils older than 50,000 years, we use Uranium 235 to measure the age of rocks the fossils are in because its half-life is about 700 million years. Why is this a more useful tool for aging very old fossils?

Types of Soil

Directions:

Use the **internet** and your **textbook** to research the seven different soil types and describe what they consist of and how they were formed.

1) Type:
    a. It consists of:

    b. It formed from:

2) Type:
    a. It consists of:

    b. It formed from:

3) Type:
    a. It consists of:

    b. It formed from:

4) Type:
    a. It consists of:

    b. It formed from:

5) Type:
    a. It consists of:

    b. It formed from:

6) Type:
    a. It consists of:

    b. It formed from:

7) Type:
    a. It consists of:

    b. It formed from:

Determining Soil Type

Directions:

You will need a **tablespoon**, a **soil sample**, a **beaker** with **water**, and a **pipette**. **Looking at the materials and lab we will be using, what are the safety precautions we should take to protect ourselves and materials during the investigation?**

1) Place a tablespoon of soil in the palm of your hand. Add water a drop at a time and knead the soil to break down all aggregates. Soil is proper consistency when moldable like putty. Use the flow diagram on the next page to find the type of soil you have by texture-by-feel- analysis.

Questions:

1) Where did you collect the soil?

2) Did the soil remain in a ball when squeezed?

3) Did it form a ribbon?

4) How long was the ribbon?

5) Did it feel more gritty, smooth, or neither?

6) What type of soil do you have?

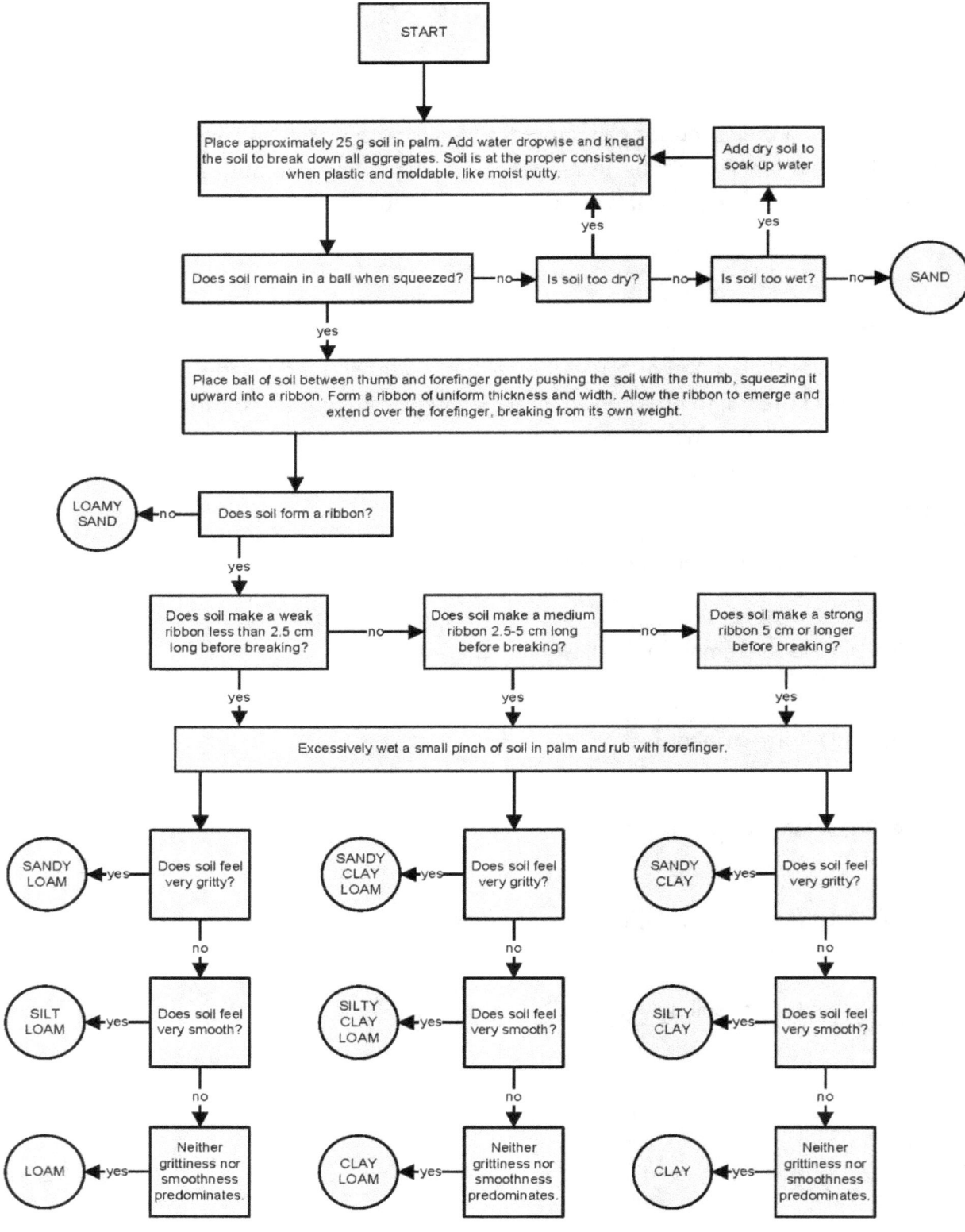

Modified from S.J. Thien. 1979. A flow diagram for teaching texture by feel analysis. Journal of Agronomic Education. 8:54-55.

Soil Salinity

Directions:

You will need a **soil sample**, a **scale**, a **250 mL beaker**, **distilled water**, a **stirring rod** or **spatula**, and a **conductivity probe** attached to an **interface** connected to a **computer** with **Logger Pro**. Looking at the materials and lab we will be using, what are the safety precautions we should take to protect ourselves and materials during the investigation?

1) Place 50 g of soil into a 250 mL beaker. Add 100 mL of distilled water and stir thoroughly.
2) Stir once every three minutes for 15 minutes.
3) Makes sure the conductivity probe unit is set to dS/m. Make sure the switch on the conductivity probe box is at 0-20000 µS/cm.
4) Place the conductivity probe into the soil and water mixture, completely covering the hole at the end of the probe.
5) Compare this to the information in Data Table 1.

Data Table 1

Salinity (dS/m)	Plant Response
0-2	Few problems
2-4	Some sensitive plants have trouble
4-8	Most plants have trouble
8-16	Only some plants will survive
Above 16	Very few plants will survive

Questions:

1) What is the salinity of the soil sample you tested?

2) According to the table, how would plants respond to the soil sample you tested?

3) Describe ways in which soil can become saline.

Soil pH

Directions:

You will need a **soil sample**, a **scale**, a **250 mL beaker**, **distilled water**, a **stirring rod** or **spatula**, and a **pH probe** attached to an **interface** connected to a **computer** with **Logger Pro**. Looking at the materials and lab we will be using, what are the safety precautions we should take to protect ourselves and materials during the investigation?

1) Place 50 g of soil into a 250 mL beaker. Add 100 mL of distilled water and stir thoroughly.
2) Stir once every three minutes for 15 minutes.
3) Rinse off the pH probe, place it into the soil mixture, and stir lightly, waiting for the pH probe to level out (this takes patience).
4) Compare this to the chart below and answer the questions that follow.

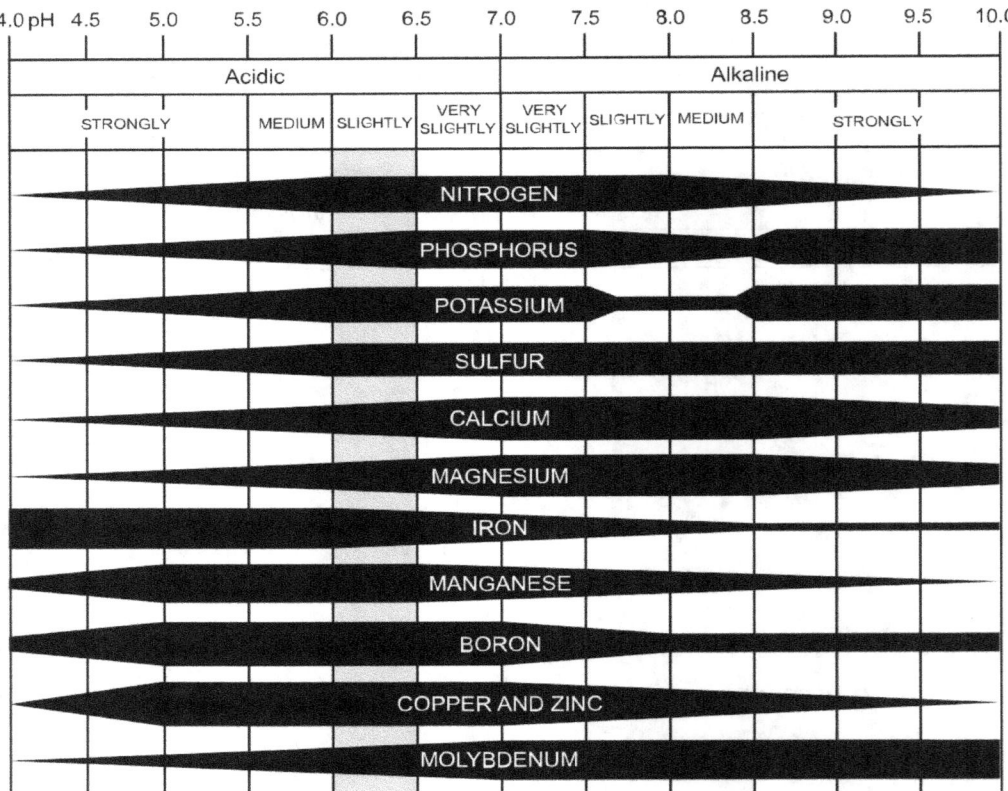

Questions:

1) What is the pH of the sample you tested?

2) Is it acidic (sour) or alkaline (sweet)?

3) What nutrients will this soil be able to hold more?

4) Which nutrients will it be able to hold less?

5) According to the chart, how is soil pH important to plants?

6) How is soil pH important in discussing acid deposition and its influence on plant growth?

Timeline of Tectonic Plate Theory

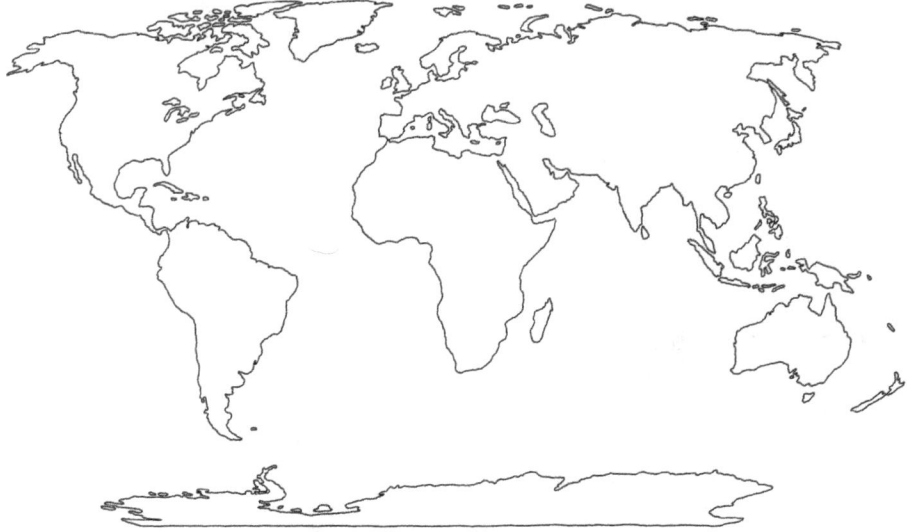

Directions and Questions:

You will use the **internet** and your **textbook** to build a timeline of the development of the modern plate tectonic theory.

1) Looking at this map, you might see what others saw when the Earth was first being mapped out. What places on this map look like they once fit together?

2) On a separate sheet of paper, starting with this idea in #1, map out how this idea progressed into the modern theory we have today.

3) Include all the evidence scientists discovered to show the Earth's plates are moving and how they are moving. What evidence did you find?

4) Construct this timeline using any media your teacher wants you to use.

Plate Tectonics Lab (Eatable)

Directions and Questions:

You will need a **clean surface**, a **large beaker of water**, graham crackers, frosting, and **fruit roll-ups**. Looking at the materials and lab we will be using, what are the safety precautions we should take to protect ourselves and materials during the investigation?

1) Clean your lab table with Soap and water. Place a glob of frosting on the table; this will represent the magma of the mantel.
2) We will do a **divergent boundary**. Take two fruit roll-ups (representing ocean plates) and lie them next to each other over the frosting. Gently press down and pull them apart. What do you see?

3) What will happen to the exposed magma?

4) So how does new ground form on the ocean floor in the Atlantic Ocean?

5) Is the Atlantic Ocean getting bigger or smaller?

6) Next, we will make **transform boundaries**. Take your two fruit roll-ups and place them next to each other again over the frosting. Now slide them along their meeting edges. What do you think this causes around the San Andreas fault in California when these plates move?

7) We will make a **convergent boundary** between an **ocean plate** and a **land plate**. Take one fruit roll-up away and replace it with a graham cracker. Move the land plate (graham cracker) into the ocean plate (fruit rollup). What do you see?

8) How do you think the islands of Japan formed?

9) Is the Pacific Ocean getting bigger or smaller?

10) Now we will make a **convergent boundary** of two **land plates**. Take your other fruit roll-up away and get another graham cracker (representing another land plate). Dip the edge of each graham cracker into the water for a few seconds, and line those edges up with each other on the frosting. Gently push down and push them together. What do you see?

11) How do you think the Himalayan Mountains are rising up between India and China?

12) How do you think heat and rock composition affect the density within the Earth's interior?

   a. How does density influence the development and motion of the Earth's tectonic plates?

13) In each of these examples, what could come through the boundary?

14) What will people near these plate boundaries feel each time they move?

Plate Tectonics Lab (Non-Eatable)

Directions and Questions:

You will need **paper** and **Play-Doh**. **Looking at the materials and lab we will be using, what are the safety precautions we should take to protect ourselves and materials during the investigation?**

1) Place a glob of Play-Doh on the table; this will represent the magma of the mantel.

2) We will do a **divergent boundary**. Take your hands (representing ocean plates) and lie them next to each other over the Play-Doh. Gently press down and pull them apart. What do you see?

3) What will happen to the exposed magma?

4) So how does new ground form on the ocean floor in the Atlantic Ocean?

5) Is the Atlantic Ocean getting bigger or smaller?

6) Next, we will make **transform boundaries**. Take your two hands and place them next to each other again over the Play-Doh. Now slide them along their meeting edges. What do you think this causes around the San Andreas fault in California when these plates move?

7) We will make a **convergent boundary** between an **ocean plate** and a **land plate**. Take one hand away and replace it with a piece of paper. Move the land plate (the paper) into the ocean plate (hand). What do you see?

8) How do you think the islands of Japan formed?

9) Is the Pacific Ocean getting bigger or smaller?

10) Now we will make a **convergent boundary** of two **land plates**. Take your other hand away and get another piece of paper (representing another land plate). Put the two pieces of paper lined up with each other and gently push down and push them together. What do you see?

   a. Another way of seeing this is to take one sheet of paper, put both hands on both sides of the paper, gently push down, bring your hands together, and see what happens to the land (paper).

11) How do you think the Himalayan Mountains are rising up between India and China?

12) How do you think heat and rock composition affect the density within the Earth's interior?

   a. How does density influence the development and motion of the Earth's tectonic plates?

13) In each of these examples, what could come through the boundary?

14) What will people near these plate boundaries feel each time they move?

Effects of Plate Tectonics

Directions:

Use the **internet** to explain how plate tectonics affects:

1) Earthquakes

2) Volcanic eruptions

3) Tsunamis

4) Mass extinctions

5) Sea level fluctuations

6) Glaciation

7) Ocean basin formation

8) Continental build up and mountain formation

9) Supervolcanoes

10) Island formation

11) Explain how and why tectonic plates move.

Calculating the Rate of Tectonic Plate Motion

Directions:

Use the map below, a **ruler**, and the formula: (island distance from the hotspot ÷ the islands age = velocity of the plate) to help you calculate the rate of movement of the ocean crust in the Pacific Ocean near the Hawaiian Islands. The hotspot is the black dot, and the triangles are where the volcanoes are on the other large islands.

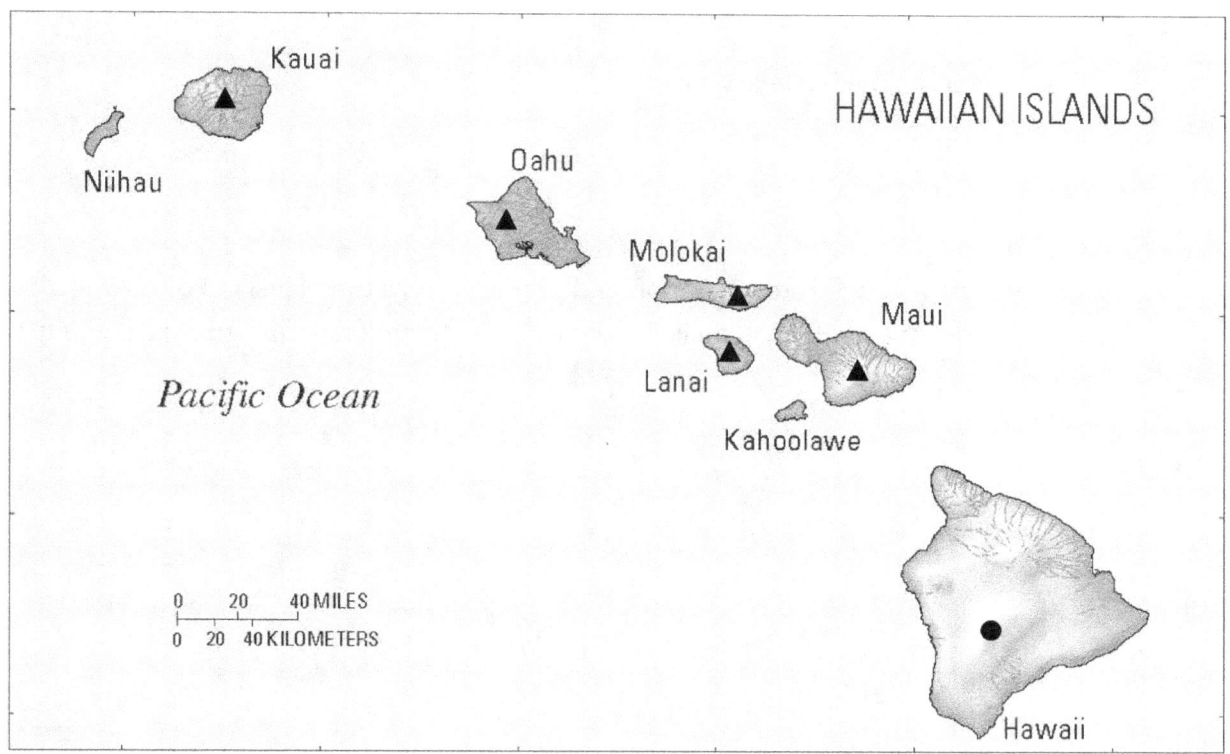

1) Calculate the rate each island has moved if the approximate age of each island is:

   a. Kauai is 5.1 million years old

   b. Oahu is 2.8 million years old

c. Molokai is 1.6 million years old

d. Lanai is 1.3 million years old

e. Maui is 1 million years old

2) Calculate the average rate of motion and predict where the big island of Hawaii would be from the hot spot in 4 million years. Place a mark on the map predicting the location.

Topographical Map of the Earth

Directions:

You will need a **topographical map** of Earth to fill out the data below.

1) Where are major mountain ranges located?

   a. Which is the tallest?

2) Approximately how far away is your home from the nearest mountain?

3) Where is the deepest part of the ocean?

4) Approximately how far away is your home from the coast of the closest ocean?

5) What different geographical features can you see on the map of the continent you live on?

   a. What type of biomes do you think are located in these different geographical features? Give evidence of how you know.

Interpreting Topographic & Geological Maps of your State

Directions:

You will need a **topographical map**, a **geological map** of your state, and a **geologic time scale** to fill out the data below.

1) What is your state?

2) Where is the highest elevation in your state?

   a. Describe the surface rocks in that area.

3) Where is the lowest elevation in your state?

   a. Describe the surface rocks in that area.

4) What are your state's different geographical features, and where are they located?

5) What patterns do you see where different types of rock are found?

   a. Why do you think the pattern is there?

6) How is a topographic map different from a geologic map?

Virtual Investigations that go with Earth's Structure and Function

ExploreLearning.com:

- Plate Tectonics

- Rock Cycle

- Rock Classification

- Mineral Identification

- Building Pangaea

Unit 5: Weather Patterns

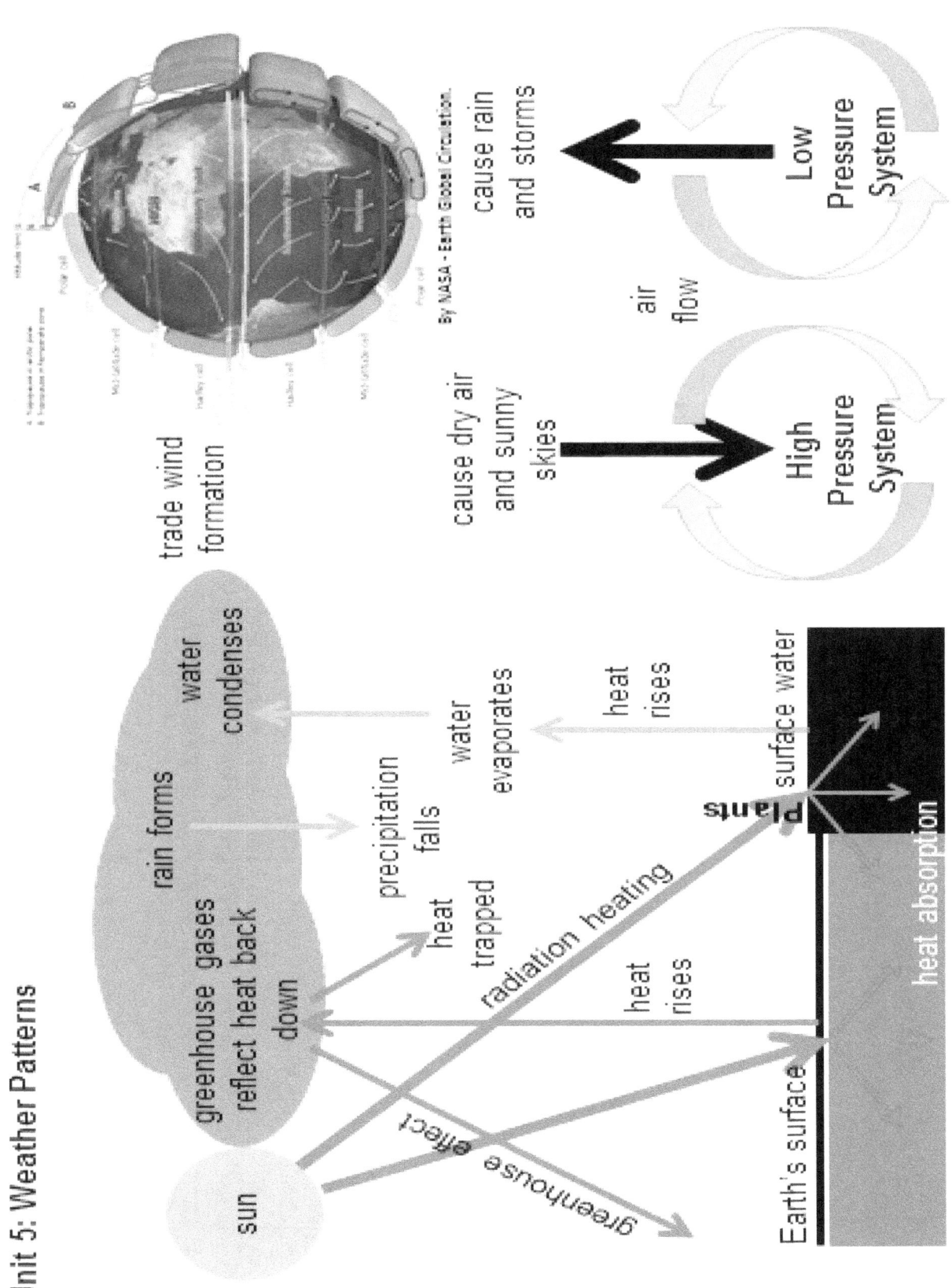

Unit 5: Weather Patterns

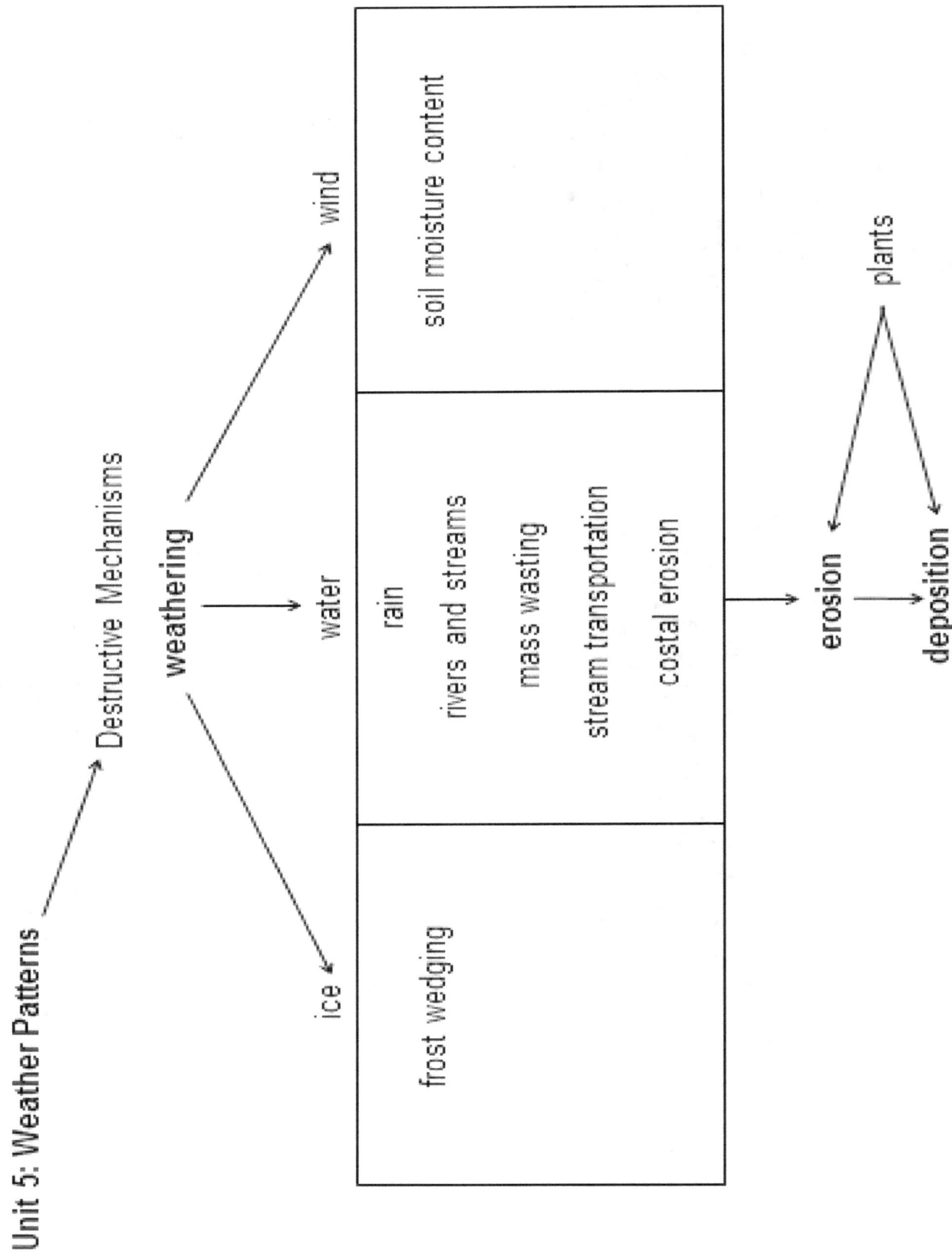

The Water Cycle (A)

Directions:

Use the **internet** or your **textbook** to draw the water cycle below. Pay close attention to what directions the water moves to make it condense into clouds and rain.

Questions:

1) What direction does water have to move to condense into clouds and rain?

2) What happens to storms if forces make air move faster in that direction?

How Does Rain Form?

Directions and Questions:

You will need a **glass** or **beaker of ice water. Looking at the materials and lab we will be using, what are the safety precautions we should take to protect ourselves and materials during the investigation?**

1) Why do you see water forming on the outside of the glass?

2) How is this like water forming droplets in the sky, making clouds and rain?

   a. What is the difference between clouds and rain?

3) How is a liquid different from a gas allowing this to happen with water?

4) What allows the water to stick together on the glass?

5) What holds the water drops to the glass?

6) How are **cohesion**, **adhesion**, and **surface tension** involved?

7) If an animal was small enough, explain how it could walk on water.

How Hurricanes Form

Directions:

Use the **internet** to go to a NASA URL address at https://tinyurl.com/4erbxnet. Use this web page to research how hurricanes form and answer the questions below.

1) Where do tropical cyclones form on Earth?

   a. Why do they form there?

2) What do hurricanes need to form?

3) Describe how a hurricane forms.

4) How does the eye form?

5) Why do hurricanes weaken when they go over land?

6) What determines the category of the hurricane?

7) Describe each category of hurricane listed here.

   a. 1

   b. 2

   c. 3

   d. 4

   e. 5

8) What causes the damage we see on land?

9) What allows us to better forecast where hurricanes will hit?

10) How do forecasters show where they predict the hurricanes will move to?

11) What may be causing the increase in the number and intensity of hurricanes?

12) What evidence did you see from NASA that this is happening?

13) What else is this causing?

14) What evidence did you see from NASA that what happens on one continent spreads to another?

15) Why is it important that all countries cooperate with each other regarding climate change?

Global Wind Movement

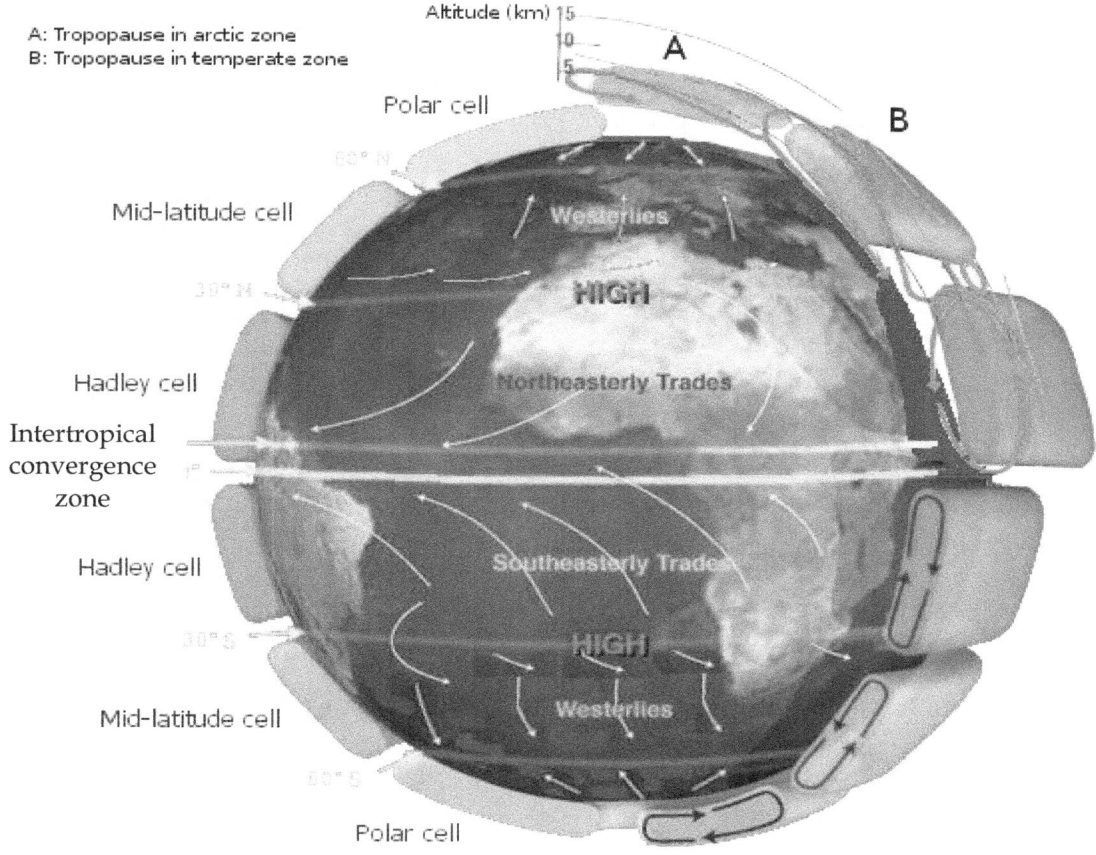

Picture by NASA - Earth Global Circulation.

Directions:

Use the NASA Picture above to help you answer the questions below on global wind patterns.

1) How many trade wind zones does the Earth have?

2) Which zone are you in on the Earth?

3) Which direction does your weather generally come from?

4) Where do the different trade winds move toward each other?

5) Where do the different trade winds move away from each other?

6) Look at the right side of the diagram. How do the convection cells change from the surface to the upper atmosphere?

7) Where is the hottest part of Earth?

8) When the sun heats the surface of this area which way does the air move (which direction does all hot air move)?

9) Heat likes to move from hot to cold areas. Where does the heat want to naturally move from the equator?

10) Using what you know about inertia, what does the atmosphere do when the Earth moves (hint: an object at rest wants to stay at rest)?

11) Which direction would this make the winds move across the Earth's surface?

12) The combination of the movement of the Earth and the direction heat wants to move gives us the trade winds; this is called the **Coriolis effect**. The Coriolis effect explains why high-pressure systems move in opposite directions in the northern hemisphere from the southern hemisphere. Why do cyclones (hurricanes) move in opposite directions in the northern and southern hemispheres?

Model Showing the Rotation of the Earth Stirring up the Atmosphere

Directions and Questions:

You will need a **spoon**, a **glass bowl** 2/3 filled with **water**, and some **pepper**. The pepper lets us better see the movement of the water when it is sprinkled in. **Looking at the materials and lab we will be using, what are the safety precautions we should take to protect ourselves and materials during the investigation?**

1) While the water is still, how does the pepper show it is moving?

2) In this model, the spoon will represent the movement of the Earth spinning on its axis, and the water will represent the atmosphere. Use the spoon to stir the water. How does the water move compared to the spoon?

3) Which is moving faster?

4) How does the pepper help you see the water moving?

5) Explain how Newton's 1st Law of Motion is involved with how the trade winds blow over the planet's surface because of the Earth's rotation.

6) What effect of the tradewinds is not included in this model?

A Local Weather Study

Directions:

You will need a **temperature probe**, a **humidity probe**, and a **UVB Sensor** attached to an **interface** connected to a **computer** with **Logger Pro**. **Looking at the materials and lab we will be using, what are the safety precautions we should take to protect ourselves and materials during the investigation?**

1) Go to a website that gives your local temperature and humidity. Each student should try a different site. We will test how accurate these sites are. Write this information in Data Table 1.
2) Take your computer, interface, and probes outside to test what the temperature and humidity actually are. Write this information in Data Table 1.
3) Subtract the difference between what the website said and what you actually measured. Write this information on the right side of Data Table 1.
4) Take the UVB sensor and point it at the sun. If possible, take measurements when there are no clouds in front of the sun. Then measure when clouds move in front of the sun. Look at the difference in the number of UV rays that come through the atmosphere with and without clouds. Write this in Data Table 1, continued on the next page.
5) Compare your results with the rest of the class to see which websites were most accurate and answer the following questions.

Data Table 1

The website used: _____	Website Measurements	Actual Measurements	Difference
Temperature			
Humidity			

UVB Reading	No Clouds:	With Clouds:	Difference:

Questions:

1) Describe the location where you recorded your measurements. Include observations such as:

   a. Is the spot open? Are there buildings, trees, or other objects that could have affected your measurements?

   b. What is the ground cover?

   c. Are there any living organisms in the immediate area?

2) Which website did the class find to be the most accurate?

3) Why do you think that website's measurements were so close to yours?

4) What do you think will happen to the humidity measurements at different times of the day?

   a. Why do you think this will happen?

5) What do you think will happen to the temperature measurements at different times of the day?

   a. Why do you think this will happen?

6) What part of the atmosphere blocks harmful UV rays from the sun?

7) Why is it important to know how many UV rays come through the atmosphere?

Relative Humidity

Directions:

You will need a cut piece of a **shoelace**, **tape**, a **beaker** of **water**, and two **temperature probes** attached to an **interface** connected to a **computer** with **Logger Pro**. Looking at the materials and lab we will be using, what are the safety precautions we should take to protect ourselves and materials during the investigation?

1) The first temperature probe will be the dry probe inserted into Channel 1 on the interface. Take the cut shoelace and tape it to the tip of the second temperature probe; insert this into Channel 2. This probe will be the wet probe.
2) Place the second temperature probe in the beaker of water to wet the shoelace.
3) Hold the first temperature probe in one hand, take the second temperature probe out of the water, and gently wave it back and forth to help the water evaporate. Wait for the temperature to even out as it drops on the second temperature probe (this could take a few minutes).
4) When the two probes have steadied their temperatures, write them in Data Table 1.
5) Repeat the procedure for #s 2-4 for two different locations of your teacher's choice.
6) Subtract the wet probe temperature from the dry probe to find the temperature differences at each place. Write this in Data Table 1.
7) Use the temperature difference with the dry probe temperature in Data Table 2 to find each humidity at the different places. Place these humidity values at the bottom of Data Table 1.

Data Table 1

	Classroom	Site 2:	Site 3:
Dry Probe Temperature (°C)			
Wet Probe Temperature (°C)			
Temperature Difference (°C)			
Relative Humidity (%)			

Data Table 2

Dry Probe Temperature	Dry Probe Minus Wet Probe Temperature(°C)									
	1	2	3	4	5	6	7	8	9	10
10°C	88	77	66	55	44	34	24	15	6	
11°C	89	78	67	56	46	36	27	18	9	
12°C	89	78	68	58	48	39	29	21	12	
13°C	89	79	69	59	50	41	32	22	15	7
14°C	90	79	70	60	51	42	34	26	18	10
15°C	90	80	71	61	53	44	36	27	20	13
16°C	90	81	71	63	54	46	38	30	23	15
17°C	90	81	72	64	55	47	40	32	25	18
18°C	91	82	73	65	57	49	41	34	27	20
19°C	91	82	74	65	58	50	43	36	29	22
20°C	91	83	74	67	59	53	46	39	32	26
21°C	91	83	75	67	60	53	46	39	32	26
22°C	92	83	76	68	61	54	47	40	34	28
23°C	92	84	76	69	62	55	48	42	36	30
24°C	92	84	77	69	62	56	49	43	37	31
25°C	92	84	77	70	63	57	50	44	39	33
26°C	92	85	78	71	64	58	51	46	40	34
27°C	92	85	78	71	65	58	51	46	40	34
28°C	93	85	78	72	65	59	53	48	42	37
29°C	93	86	79	72	66	60	54	49	43	38
30°C	93	86	79	73	67	61	55	50	44	39

Questions:

1) How did the wet probe temperature compare with the dry probe temperature at each site? Explain why.

2) Which site had the highest relative humidity?

3) Which site had the lowest relative humidity?

4) Explain what could cause the relative humidity to be different at these locations on the same day.

5) How do you think the humidity would change during the day at different times outside?

6) Compare the relative humidity values outside at sunny and shaded sites under trees.

7) What could be some sources of error in this investigation?

Temperature Inversions

Directions:

Use the **internet** to research temperature inversions, then answer the following questions.

1) What are temperature inversions?

2) What are some examples of temperature inversions, and where can we find them?

3) What are the short-term effects of temperature inversions?

4) What are some long-term effects of temperature inversions?

5) How do they affect El Nino and La Nina oscillations?

6) How do they affect the polar ice caps and glacial melting?

7) How do they change the ocean surface temperatures?

Weathering

Directions and Questions:

You will need **sugar cubes** and **chalk** to represent rocks, **water** in a **beaker**, **vinegar** in another, **pipettes,** and a small **tray/pan.** Weathering is the breakdown of rock by wind, water, and chemical reactions. **Looking at the materials and lab we will be using, what are the safety precautions we should take to protect ourselves and materials during the investigation?**

1) The wind has particles that, when they hit a rock, microscopic pieces break off, making the rock smaller. Rub your hand across a sugar cube while it is over the tray/pan; what does this do to the cube?

2) Take another sugar cube, place it in the tray/pan, and squirt water on it from a pipette. What does this do to the sugar cube?

3) What happens to water when it freezes?

    a. How can this action help break apart rocks?

4) Take some chalk (representing limestone) and squirt water on it. Now take some vinegar (representing an acid) and squirt that on the chalk. What do you see chemically taking place?

5) How does this lab model weathering?

6) What is not accurate about this model?

Erosion

Directions:
You will need an **apron, safety goggles, shovel, dirt** or **sand** in a **large tub/pan**, a **pitcher** of **water**, a **hairdryer**, a **spray bottle** that sprays **water**, the **internet,** and a **textbook. Looking at the materials and lab we will be using, what are the safety precautions we should take to protect ourselves and materials during the investigation?**

1) Go outside, dig up some dirt/sand, and put it into your large tub/pan. Fill it half full. Bring it inside to your lab table.

2) Angle your tub/pan so one end is higher than the other. We are going to simulate **wind** hitting the ground with a hairdryer. Put on your safety goggles. Plug in your hair dryer and keep it away from water. Turn on your hairdryer, have everyone in your group stand behind it so dirt does not blow on them, and bring the hairdryer closer to the dirt in your tub. What do you see happening to the small particles of dirt?

    a. How would plant life rooted in the dirt and covering it affect this action?

    b. You can place your hand in front of the blowing dryer to see. How does putting an object between the wind source and the dirt affect the movement of the dirt?

3) Unplug your hairdryer and set it aside. Take your spray bottle to simulate what a light rain might do to the dirt. Set it to mist and spray the dirt. This model simulates **sheet erosion**. How does the dirt move? Describe the pattern it is making.

4) Now turn on the hairdryer again and aim it at the moist dirt. How does soil with moisture behave differently in the wind than soil without moisture?

5) Adjust your spray bottle to make a stream come out. This model will simulate heavier rain. Aim your bottle at an area of the dirt standing behind the high end of the tray and spray. How is the dirt movement different?

   a. What do you see forming in the dirt?

   b. Have multiple spray bottles spray at the same time. This model will simulate **rill erosion**. How is this pattern different than sheet erosion?

   c. How do you think plant life rooted in the dirt and covering it affects this action?

6) Now you will take your pitcher of water to simulate **gully erosion** and pour it from the high end of the tub. How is this dirt movement different from the others?

7) Take your wet dirt/sand back outside and dump it where your teacher tells you to. And dry out your tub/pan for the next class.

8) Use the internet and your textbook to fill in the chart below describing the three types of erosion we just modeled. Then research answers to the questions that follow.

Chart 1

Type	Picture	Description
Sheet Erosion		
Rill Erosion		
Gully Erosion		

Questions:

1) How does plant life slow the erosion of soil?

2) Why do most yards have grass on them?

3) What happens when that grass is worn away?

4) Fill in the chart below describing how we slow the erosion rate with each method.

Chart 2

Method	Description
Contour Plowing	
Strip-Cropping	
Terracing	
Crop Rotation	
Windbreaks	

5) Describe how conservation tillage reduces the amount of soil erosion.

6) How do cover crops and after-harvest practices protect soil from erosion?

Biome Research Report

Have each student use the **internet** to fill these charts out for a different biome so all biomes will be researched in the class.

Ecosystem	
Location	
Climate	
Temperature include degrees °F (°C)	
Precipitation amount and pattern	
Soil description	
Plants and other producers	
Animals	
Additional Notes • human impact • interesting facts • connections	

Keystone Animals in my Biome

Notes: Be sure to indicate if the species is common, an endangered species, indicator species, keystone species, or introduced species.

Animal: Range: if migratory give summer and winter range Habitat description: Diet: Other: including interactions with other species.	**Animal:** Range: if migratory give summer and winter range Habitat description: Diet: Other: including interactions with other species.	**Animal:** Range: if migratory give summer and winter range Habitat description: Diet: Other: including interactions with other species.
Animal: Range: if migratory give summer and winter range Habitat description: Diet: Other: including interactions with other species.	**Animal:** Range: if migratory give summer and winter range Habitat description: Diet: Other: including interactions with other species.	**Animal:** Range: if migratory give summer and winter range Habitat description: Diet: Other: including interactions with other species.

Keystone Plants in my Biome

Notes: Be sure to indicate if the species is common, an endangered species, or introduced species. You may also want to note if it is an indicator species or a keystone species.

Plant:	**Plant:**	**Plant:**
Description:	Description:	Description:
Habitat description:	Habitat description:	Habitat description:
Importance: including interactions with other species.	Importance: including interactions with other species.	Importance: including interactions with other species.
Plant:	**Plant:**	**Plant:**
Description:	Description:	Description:
Habitat description:	Habitat description:	Habitat description:
Importance: including interactions with other species.	Importance: including interactions with other species.	Importance: including interactions with other species.

Biomes Chart

Use the **internet** to fill in the chart for the characteristics of each biome. Then shade in the maps that follow to show where each biome is located on Earth.

Biome	General Description	Plant Adaptations	Animal Adaptations	Threats
Tundra				
Temperate Coniferous Forest (Boreal Forest/Taiga)				
Temperate Deciduous Forest				
Grasslands				

Biomes Chart Continued

Biome	General Description	Plant Adaptations	Animal Adaptations	Threats
Dessert				
Chaparral				
Tropical Savana				
Tropical Rain Forest				

Color in where each Biome is found on the Earth

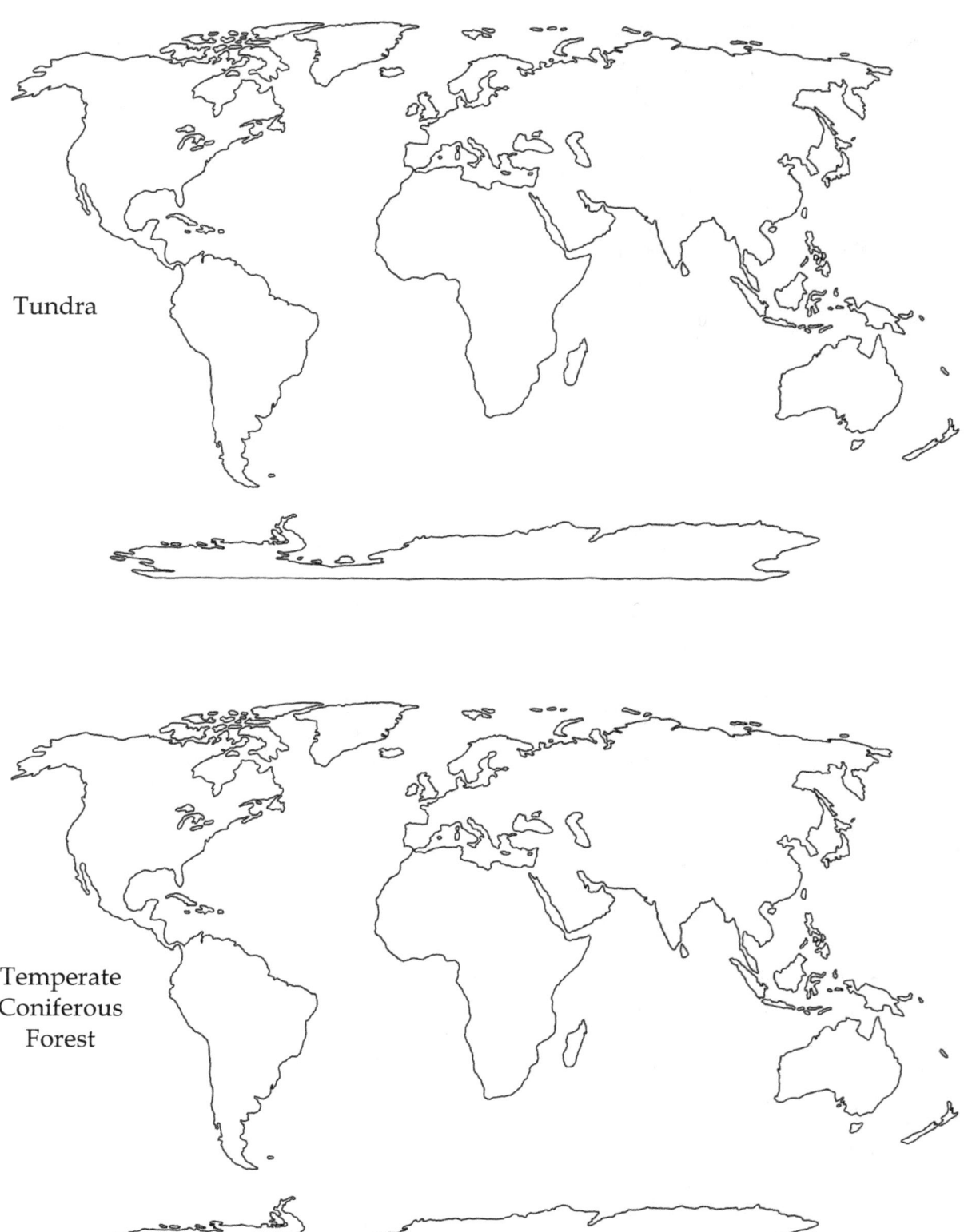

Tundra

Temperate Coniferous Forest

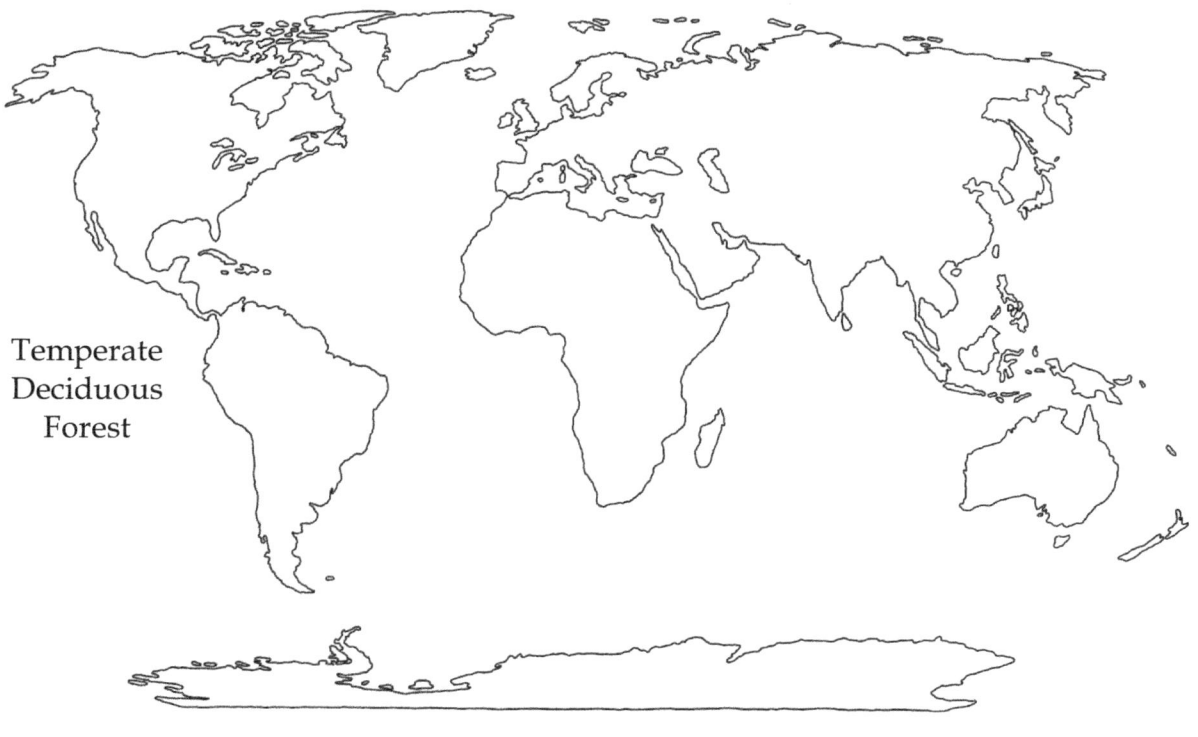

Temperate Deciduous Forest

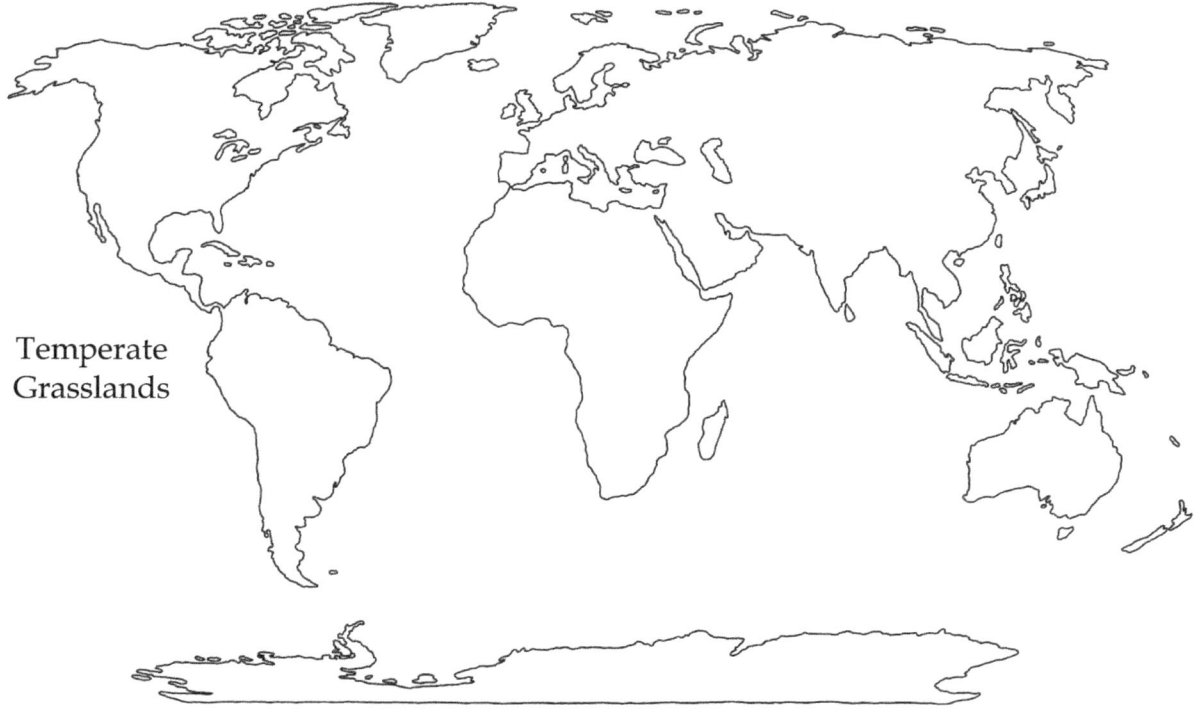

Temperate Grasslands

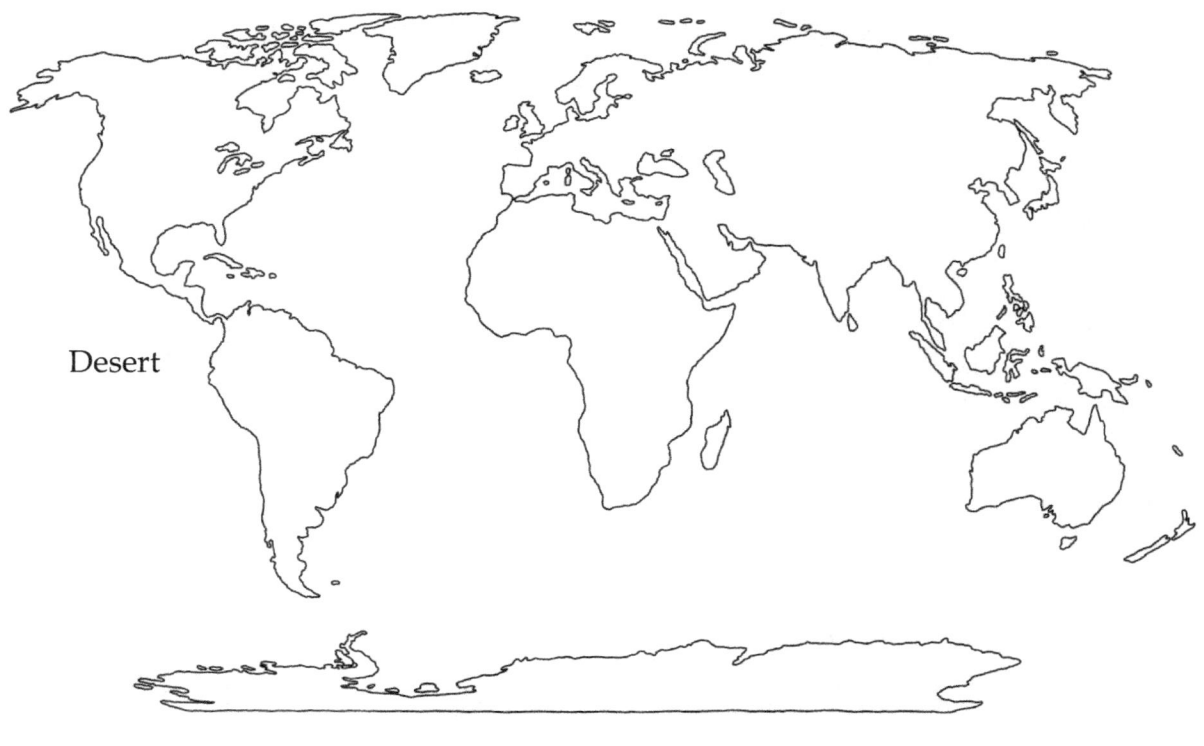

Desert

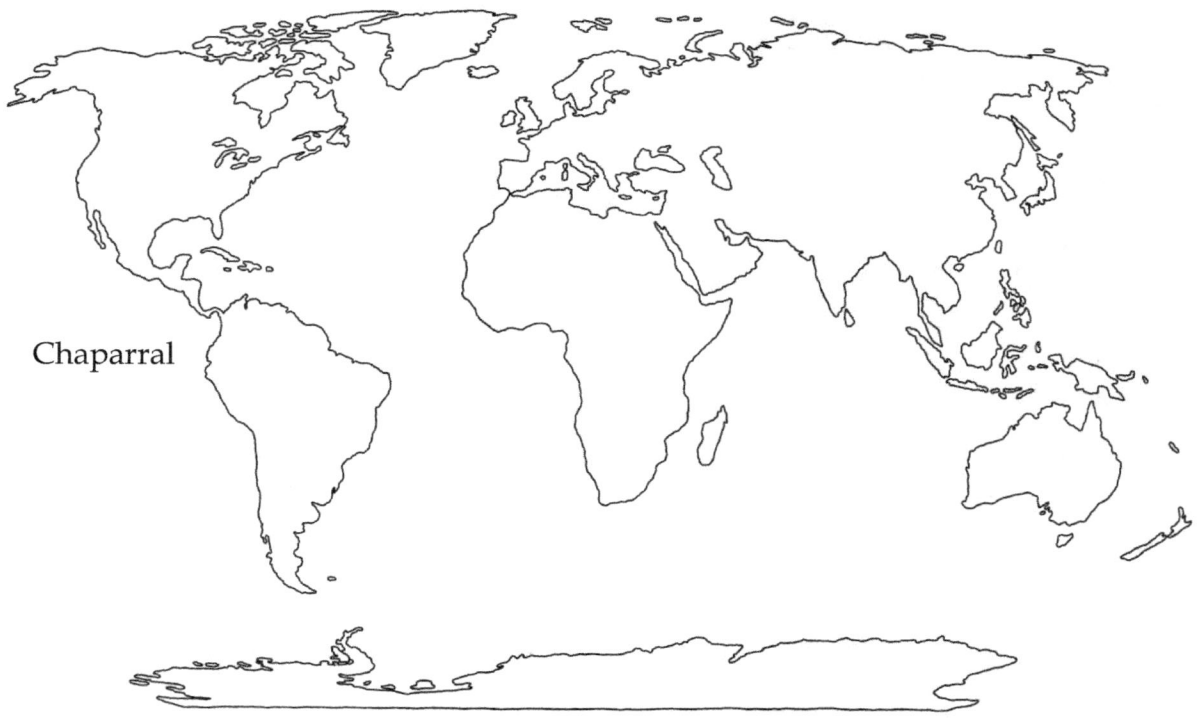

Chaparral

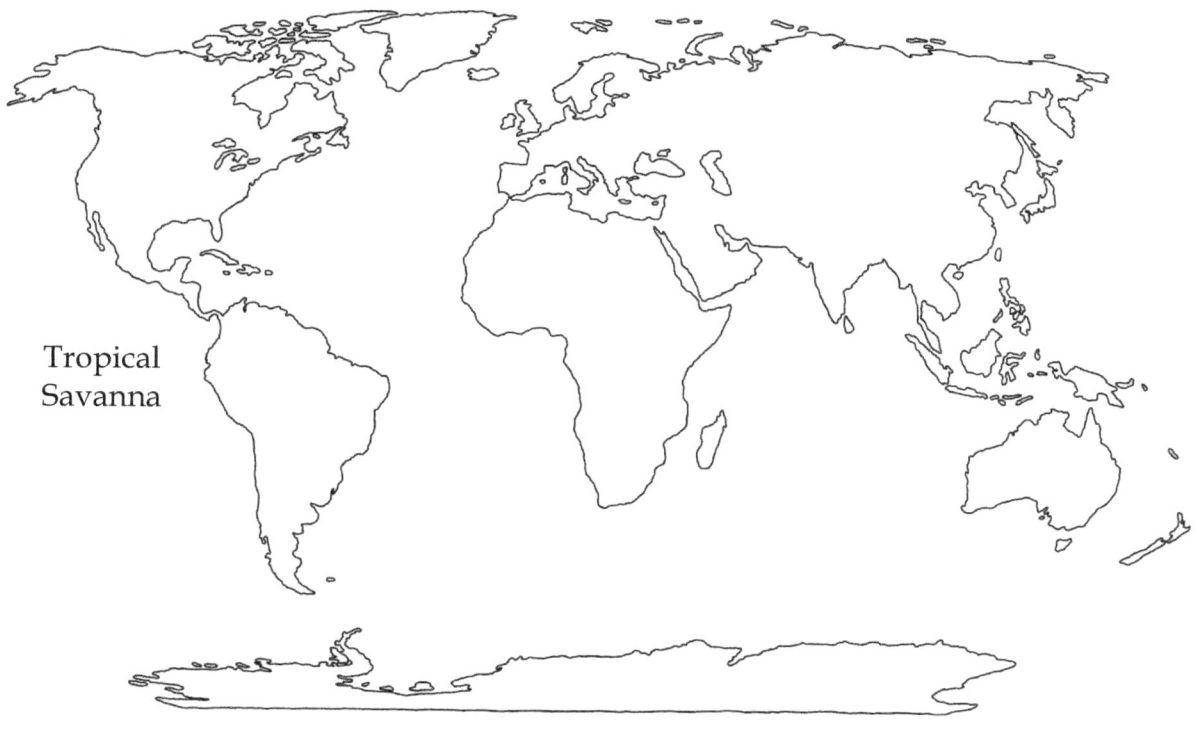

Tropical Savanna

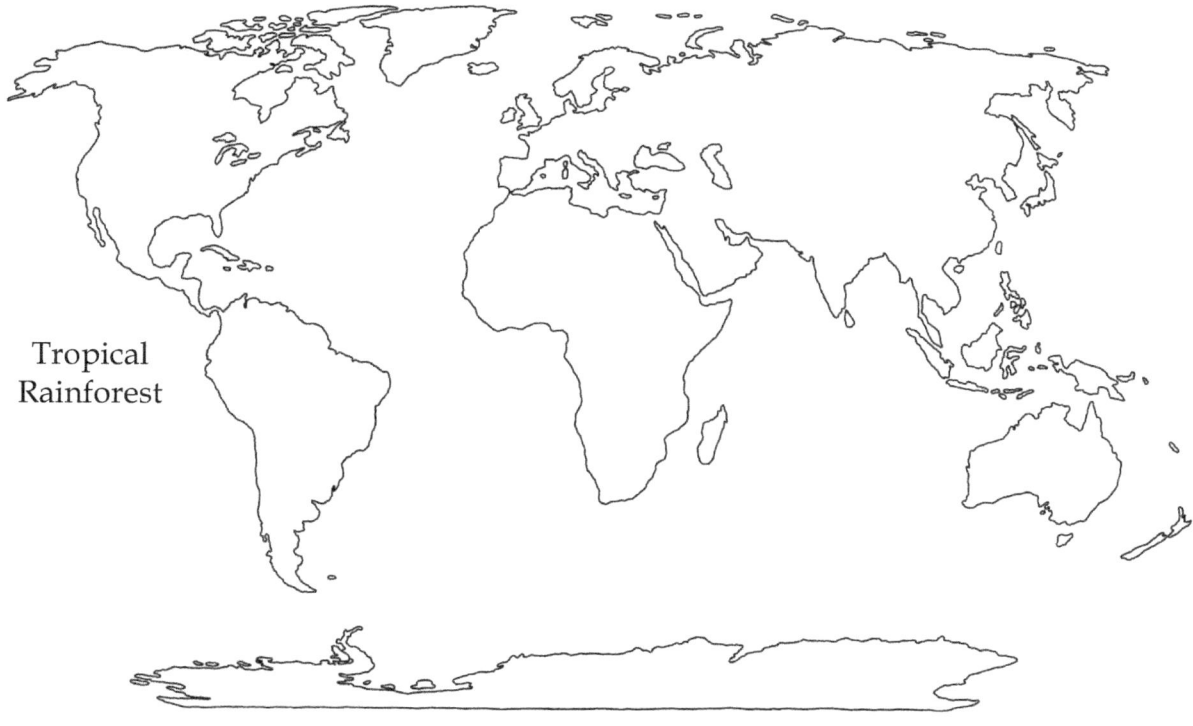

Tropical Rainforest

Virtual Investigations that go with Weather Patterns

ExploreLearning.com

    Weather Maps – Metric Gizmo

    Relative Humidity Gizmo

    Coastal Winds and Clouds Gizmo

    Weather Maps

    Observing Weather (Metric) Gizmo

    Observing Weather (Customary) Gizmo

    Comparing Climates (Metric) Gizmo

    Comparing Climates (Customary) Gizmo

    Hurricane Motion Gizmo

    Seasons Around the World

    Seasons in 3D

    Seasons: Why do we have them?

    Summer and Winter

    Water Cycle

    Convection Cells

    Greenhouse Effect

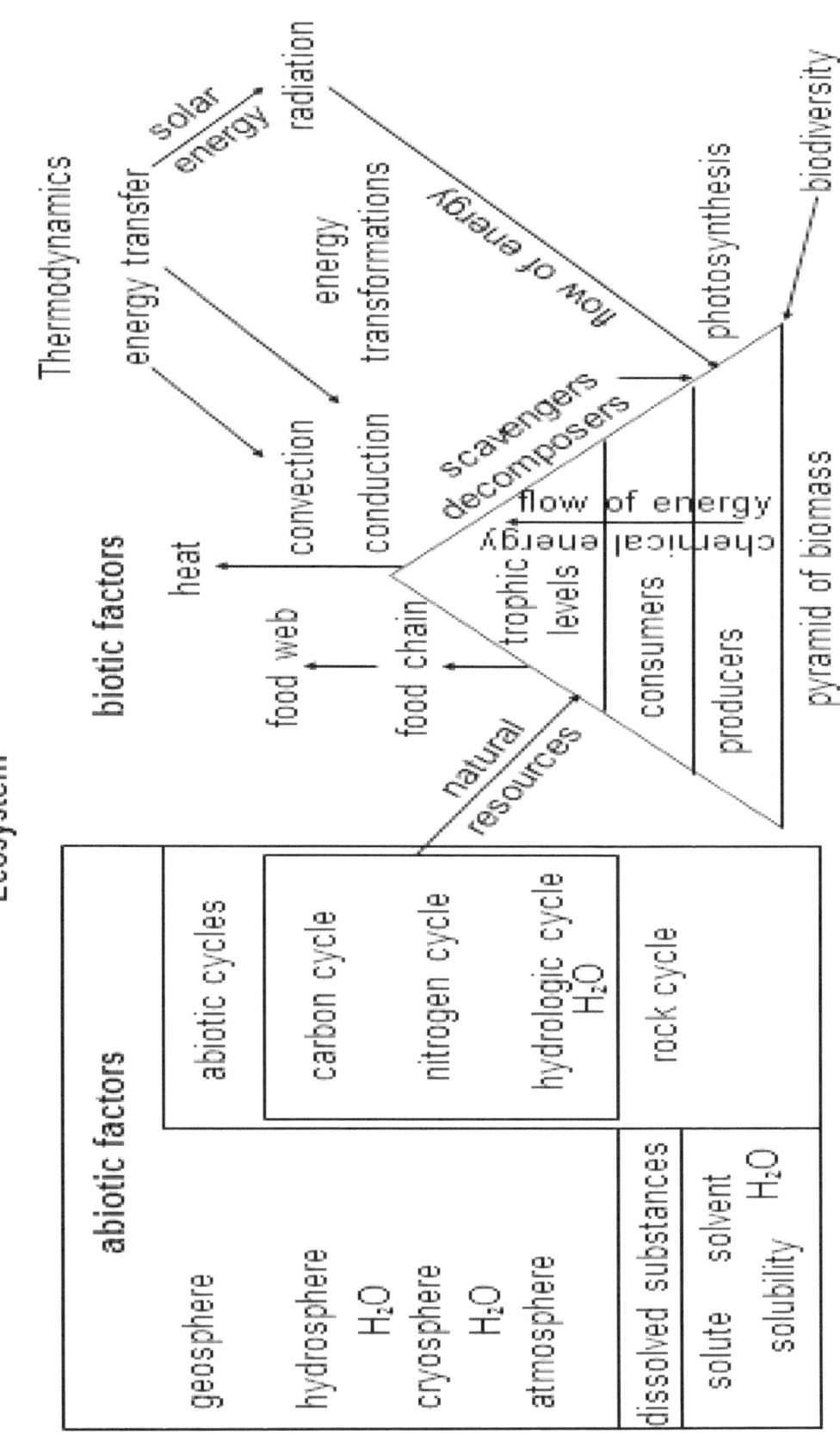

Build an Ecosystem

Directions:
We are going to build a model ecosystem out of **Play-Doh**. Use the 10% rule to help you find the amount of Play-do you will need for each trophic level. Measure the amount of biomass with a **scale** for each level in your community (see the chart below). Then form the plants and animals that will be living in your community. Because the number will be so small, when you get to the quaternary consumer (red), it will be best for the teacher to make one animal for the whole class to migrate between all the small communities. After each group has built its community, the whole class will represent a larger community (the biotic parts of an ecosystem). **Looking at the materials and lab we will be using, what are the safety precautions we should take to protect ourselves and materials during the investigation?**

Data Table 1 (Group Data)

Play-Doh Color	Trophic Level	Amount Used (g)
Green	Producer (Plants)	200.00g
Blue	Primary Consumer (Herbivore)	g
Orange	Secondary Consumer (1st Level Carnivore)	g
Purple	Tertiary Consumer (2nd Level Carnivore)	g
Red	Quaternary Consumer (3rd Level Carnivore)	g
Black	Scavengers/Decomposers	5.00g

Data Table 2 (Whole Class Data)

Play-Doh Color	Trophic Level	Amount Used (g)
Green	Producer (Plants)	g
Blue	Primary Consumer (Herbivore)	g
Orange	Secondary Consumer (1st Level Carnivore)	g
Purple	Tertiary Consumer (2nd Level Carnivore)	g
Red	Quaternary Consumer (3rd Level Carnivore)	g
Black	Scavengers/Decomposers	g

Questions:

1) Why did we use the amounts we did for each tropic level?

2) Why are high-level predators very territorial?

3) Compare and contrast the pyramid of biomass, energy, and the number of organisms.

4) What do you think determines how many decomposers an ecosystem will have?

5) Can an animal occupy more than one tropic level simultaneously in an ecosystem? Explain.

- Clean up by putting all the Play-Doh back in their original containers.

Making a Food Web

Directions:

You will need large **butcher paper, scissors, glue, colored pencils**, a **ruler**, and a **meter stick**. **Looking at the materials and lab we will be using, what are the safety precautions we should take to protect ourselves and materials during the investigation?**

1) With colored pencils, mark each group a different color. Keep in mind many organisms will have more than one color.
    a. Energy source – yellow
    b. Producers – green
    c. Herbivores – blue
    d. Carnivores – red
    e. Omnivores – orange
    f. Scavengers – purple
    g. Decomposers – brown

2) Use scissors to cut out the pictures. According to trophic levels, sort the pictures into groups: producers, herbivores, first-level carnivores, second-level carnivores, scavengers, and decomposers. Try to put the organisms near their food and predators (producers near the sun, insects in one area, and rodents in another). On some butcher paper, spread the organisms apart and give space to draw your arrows to see who eats who. Use arrows to point away from the food source and point to who is doing the eating.

3) Look at your food web; organisms with a star on them were sprayed with the insecticide DDT or ate an organism with it. If animals eat the sprayed organisms, they take in the poison. The organisms may not die, but the poison builds up in the organs of its body. Because predators eat more food that may be affected by the poison, more poison is concentrated in the higher-level consumers. **Place a red square** around the organisms in the food web that might get some poisonous DDT into their bodies from their food.

This page will be cut from the previous page.

Ecological Pyramid

Instructions:

Side 1: Label food chain with the names of organisms.
Side 2: Draw pictures of the organisms.
Side 3: Label Trophic Levels: Producers, primary consumers, secondary consumers, ect.
Side 4: Starting with 14,800 Kcal, show how the energy flows with the 10% rule. Put your name on it, cut out the model by cutting along the outer edge, fold the corners and flaps to make a pyramid. Add paste or glue to the flaps to hold it together.

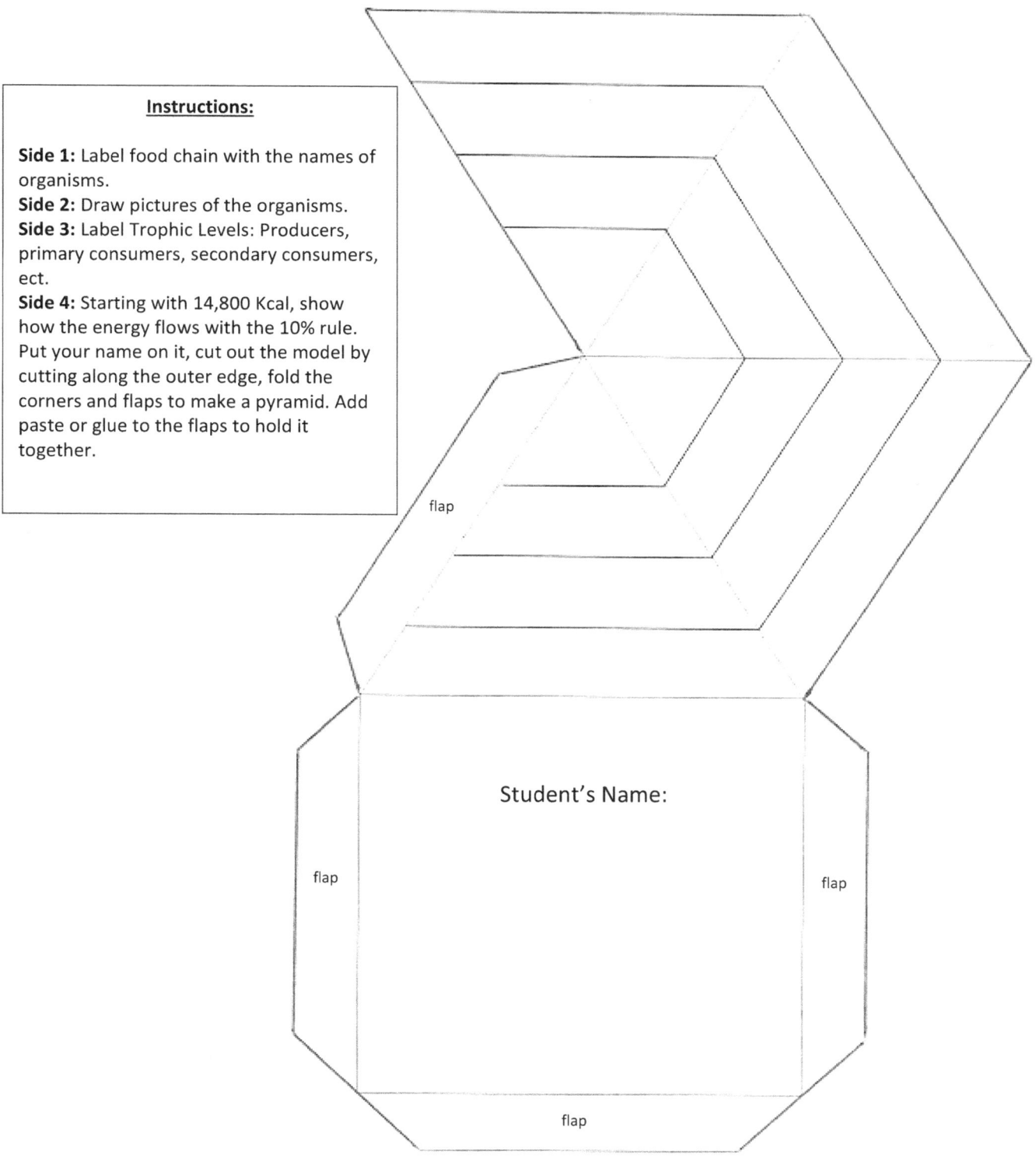

Student's Name:

This page will be cut from the previous page.

Symbiotic Relationships

Directions and Questions:

1) Read how the different types of symbiosis (the interaction relationship between organisms) can be defined.
    a. **Mutualism** can be defined as two organisms interacting where both benefit. **+ +**
    b. **Commensalism** can be defined as two organisms interacting where one benefits and the other is neither harmed nor benefits. **+ o**
    c. **Parasitism** can be defined as two organisms interacting where one benefits and the other is harmed. **+ −**
    d. **Predation** can be defined as where one benefits by eating the other. **+ −**
2) Use these definitions to classify relationships between organisms interacting below and tell why they have that type of relationship.
    a. A liver fluke causes disease in an animal.

    b. Lichens are a combination of organisms where the fungus absorbs minerals and give them to algae, and the algae go through photosynthesis to make sugar for energy and give it to the fungus.

    c. A Remora eats the scraps left behind from the feeding of a shark. The shark receives nothing in return.

    d. A dog gets fed, a home, socialization with a family of humans. The humans get home security, lower blood pressure, the release of stress, entertainment, companionship, and a sense of importance when taking care of the dog.

    e. A robin chases and eats a cicada.

    f. A cow eats grass, and the grass gets digested by the bacteria living in the cow's digestive tract releasing the nutrients.

    g. A bird eats the parasites in the mouth of an alligator.

h. Ticks and fleas live on a dog, eating its blood, thus causing the dog to itch.

i. A clownfish lives within the sea anemone. The fish gets protection and food, and the anemone gets rid of parasites as the fish eats them.

j. Mistletoe gets food and water from the tree, and the tree gets damaged.

k. A bird makes a nest in the tree, and the tree does not get helped or harmed.

l. Tardigrades are microscopic animals that pierce plant cells and suck the contents out through their straw-like mouths.

3) Go outside and find an example of each you observe and write it below.
    a. Mutualism

    b. Commensalism

    c. Parasitism

    d. Predation

Competitive Relationships

Directions:

Use the picture below to identify the biotic and abiotic factors in a model ecosystem. Then discuss with your teacher and class the questions that follow and write down your answers.

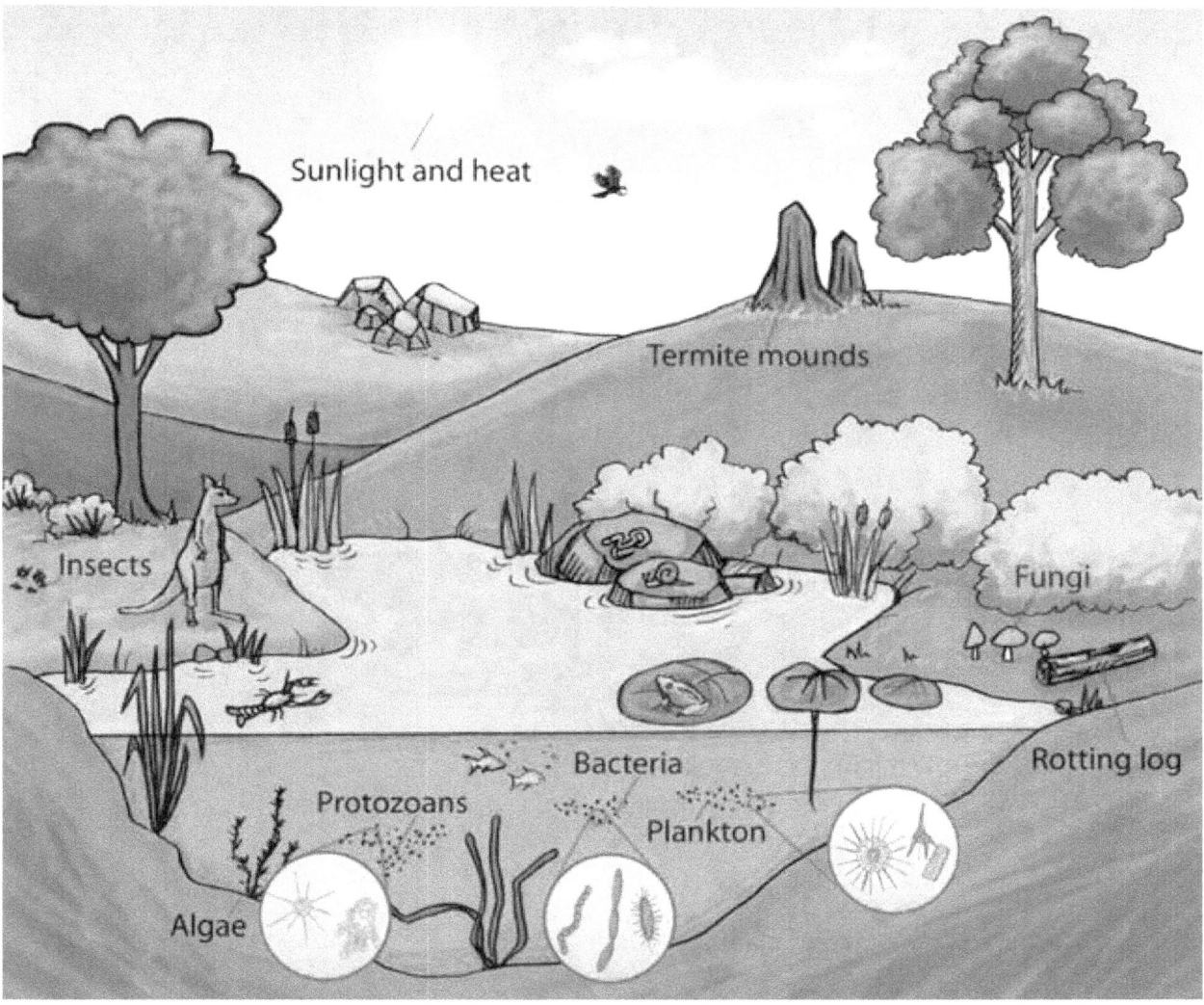

Picture from acamrmichael.weebly.com

Questions:

1) What are the biotic factors in this ecosystem?

2) What are the abiotic factors in this ecosystem?

3) Which organisms would compete with each other?

4) What would they be competing for?

5) Which organisms here would be competing for soil?

   a. How might they compete for this soil?

   b. How might plants avoid this competition?

6) Where might you find plants competing for light?

   a. How would they compete for the limited light available?

   b. How might plants avoid this competition?

7) Where might we find plants and animals competing for water?

   a. How would they compete for limited water opportunities?

   b. How might plants avoid this competition?

8) How do animals compete for food?

   a. How might animals avoid competition for a food source?

9) How could animals compete for a range of temperatures?

   a. How might animals avoid competition for a temperature range?

10) How could plants compete for a range in temperature?

   a. How might plants avoid competition for a temperature range?

11) Why do you think no two species can occupy the same niche for a long period of time?

12) What happens when an invasive species enters a new ecosystem it was not in before?

Hierarchical Organization of Ecosystems

Directions:

1) Suppose each shape represents a different species in an ecosystem. Use the picture on the next page as you follow the directions and answer the questions.
2) Circle one organism of each species.
3) Count the population of each species and write it in Data Table 1.

Questions:

1) How many populations are in this community?

2) Which part of the ecosystem is this model missing? Give examples of what some are.

3) Which shape is probably the producer? Explain why.

4) Which shape is probably the primary consumer? Explain why.

5) Which shape is probably the first-level carnivore? Explain why.

6) Which shape is probably the top of the food chain (the highest level carnivore)? Explain why.

7) Explain the relationships between an organism, population, and community in an ecosystem.

Data Tale 1:

Species	▲	■	●	⬟
Population				

Conservation of Life: Photosynthesis and Respiration

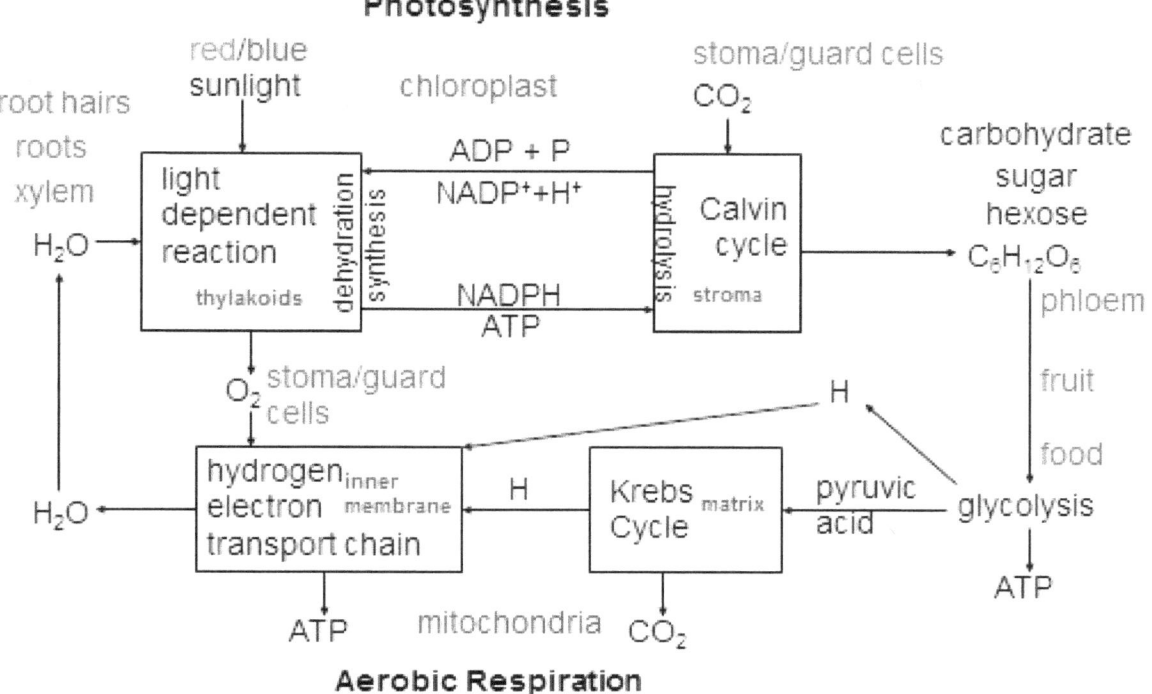

Directions:

Use the diagram above to answer the following questions.

1) Write the equation for photosynthesis and balance the reaction.

2) Write the equation for aerobic respiration and balance the reaction.

3) In both equations, trace where each element of the reactants go to make the products.

4) How are the two reactions similar to each other?

5) How does the conservation of mass, in this case, show that life has balance?

6) Plants and algae go through both photosynthesis and respiration. Animals only go through respiration. What would happen to life on Earth if we lost the plants and algae?

7) Keeping this in mind, which do you think formed first: the process of aerobic respiration or photosynthesis? Explain why.

8) How was this diagram a good model showing the conservation of mass in photosynthesis and respiration?

9) What careers would use the information shown by this activity?

Biodiversity in Ecosystems

Directions:

Use the **internet** to fill in the chart below to find and rank the biodiversity of each biome. Then answer the questions that follow.

Biome	General Description	Plant Diversity # of Species	Animal Diversity # of Species	Diversity Rank
Tundra				
Temperate Forest				
Desert				
Grasslands				
Tropical Rain Forest				

Questions:

1) Which of the Biomes in the chart showed the greatest biodiversity?

2) Why do you think there is more life there?

3) Which of the biomes showed the least diversity?

4) Why do you think there are fewer types of life there?

5) How does biodiversity show us an area's ability to support life?

6) If there is a change in the environment, which is more likely to survive the change, an ecosystem with high biodiversity or an ecosystem with low biodiversity? Explain why.

7) Does biodiversity increase or decrease around human populations?

8) Why do you think this happens?

9) When humans grow crops, how many species are planted in one field?

10) Does this increase or decrease biodiversity?

11) Where do you think there is greater biodiversity in an energy pyramid (in the producers, herbivores, or carnivores)? Explain.

12) How does diversity help support an ecosystem?

13) Which ecosystem would be easier to destroy, one with high biodiversity or one with low biodiversity? Explain Why.

Virtual Investigations that go with Energy Flow and Cycles

ExploreLearning.com:

Cell Energy Cycle

Photosynthesis Lab

Plants and Snails

Food Chain

Coral Reefs 1 – Abiotic Factors

Coral Reefs 2 – Biotic Factors

Water Cycle

Carbon Cycle

Ecosystems STEM Case

Ecosystems Handbook

Ocean Carbon Equilibrium STEM Case

Ocean Carbon Equilibrium Handbook

Photosynthesis STEM Case

Photosynthesis Handbook

Nitrogen Cycle STEM Case

Nitrogen Cycle Handbook

Cell Respiration STEM Case

Cell Respiration Handbook

Prairie Ecosystem

Forest Ecosystem

Pond Ecosystem

Unit 7: Ecological Changes

Concept map (terms as shown):

- Evolution → natural selection
- natural selection → mutations, gene flow, sexual reproduction, **recombination of genes**
- recombination of genes → homologies
- recombination of genes → **populations:** frequency of anatomical traits, variation in populations, diversity
- populations → adaptation → **selective pressures:** environmental resources, ecosystem interactions
- selective pressures → speciation, evolve, change in fitness, extinction, stasis
- selective pressures → convergent evolution, divergent evolution, co-evolution
- speciation → (energy pyramid: producers, consumers, trophic levels; chemical energy; flow of energy)
- succession, invasive species → energy pyramid
- energy pyramid → food chain → food web

Unit 7: Ecological Changes

Succession

Primary Succession	Both	Secondary Succession	
Caused by	Happen by	Caused by	
Nature	Nature	Nature	Human influence
volcanic eruption	pioneer community → change → climax community	natural disasters	mining
glaciers		hurricanes	fires
	habitats / populations / communities	flood	deforestation
Land:		fires & deforestation	
Aquatic:		mudslides / avalanche	municipal development
		tsunami	population growth
starts on rock	feedback loops	starts on soil	

regional changes ⟶ environment ⟶ global effect

The Story of Life

Directions:

In some form or fashion, build a scale model of our history of life. Make the time equal to some standard length on a **meter stick**. Use the information below that we got from the fossil record to trace our lineage and build your Story of Life Time-line:

4.6 billion years ago, Earth formed

3.8 billion years ago, first life appeared

3.4 billion years ago, photosynthesis appeared

2.9 billion years ago, aerobic respiration appeared

2.0 billion years ago, protists appeared

1.0 billion years ago, primitive plants appeared

900 million years ago, fungi and simple animals appeared

700 million years ago, sexual reproduction appeared (the ability of life to mix genes speeding up evolution)

540 million years ago, oxygen levels rose, triggering the Cambrian explosion

530 million years ago, fish appeared

410 million years ago, amphibians appeared

360 million years ago, reptiles appeared

250 million years ago, temperatures rose from volcanic activity, and dinosaurs appeared

213 million years ago, mammals and birds appeared

200 million years ago, dinosaurs became the dominant creatures

65 million years ago, dinosaurs disappeared after an asteroid impact, triggering an ice age, and mammals became dominant

55 million years ago, primates appeared

7 million years ago, apes appeared

4 million years ago, hominids appeared

1 million to 100,000 years ago, Homo sapiens appeared

Questions:

1) How does this model show the Theory of Evolution?

   a. Why is it not a hypothesis?

   b. Why is this not a law?

2) What do you think caused life to change on Earth when looking at this timeline?

3) How do you think the formation of Pangaea helped amphibians appear 410 million years ago?

4) If placental mammals did not exist in Australia but existed everywhere else on Earth, what does that tell us about how Pangaea started to break apart?

5) The fossil record shows that 99.9 % of all species that have ever lived on this Earth are now extinct. Why are they extinct, and why is the Earth not empty of life now?

Fossil Evidence of Relative Dating

Directions:

Use the diagram below modeling rock layers with a simple summary of different fossils. **Relative dating** is how scientists look at the history of life to see when things lived relative to each other. When we look at the rock layers, we find the oldest at the bottom and the youngest at the top. Use this diagram to explore this concept and answer the questions that follow.

Cenozoic Era	Hominids showed up Dominated by Mammals and Birds	A
Mesozoic Era	Dominated by Reptiles	B
Paleozoic Era	Dominated by Fish	C
Precambrian Era	Multicellular Invertebrates showed up	D
	Single Cell Eukaryotes showed up Single Cell Bacteria showed up	E
	No life found	F

Questions:

1) What kind of life was the first to show up?

2) Which kind of life was the last to show up?

3) 99.9% of all the species of life on this Earth are now extinct. Is this Earth empty of life?

   a. Could all the life live together at the same time?

b. So how could all this life have lived on this earth?

4) Which layer is the oldest?

5) Which layer is the youngest?

6) Mammals were around in the Mesozoic era; what do you think happened for them to be popular and dominate the Earth in the Cenozoic Era?

7) Describe the complexity of life as it formed through the history of the Earth.

8) Why are the older fossils found under the younger fossils?

9) If different types of living things lived at different times, what does that say about the ecosystems?

   a. What does that say about the abiotic factors that affected those ecosystems?

10) The one thing that never changes on this Earth is that everything changes. Will this Earth look the same as today in a billion years? Explain why.

Use the internet to explain:

11) How transitional fossils show lineages of life.

        i. What are some examples?

12) How evidence of biozones and faunal succession in rock layers reveal information about the environment at the time rocks were formed.

13) How do sedimentation, fossilization, and speciation affect the degree of completeness of the fossil record?

Variation Within a Population

Directions:

Each student will need ten **leaves** of any kind (must all be of the same species), ten **shelled nuts** or **seeds** (all of the same species) of any kind, and a **metric ruler**. The more students you have, the bigger your data set, and the better results you will get. **Looking at the materials and lab we will be using, what are the safety precautions we should take to protect ourselves and materials during the investigation?**

1) Go out to a tree and randomly take ten leaves off the tree. When you take the leaf off, make sure you do not rip any part of the leaf. Measure the length of the longest part of the leaf in millimeters and write this data in Data Table 1.
2) Randomly take ten whole-shelled nuts or seeds out of a bag. Do not use the ones that are broken. Measure the length in millimeters and write this data in Data Table 1.
3) Find the measurement of the shortest leaf in the class and the longest leaf in the class. Then fill in the equal increments between those measurements to make 14 different groups. Then take a class count of how many leaves fit in each of those categories. Write this data in Data Table 2.
4) Find the shortest nut pod/seed and the longest nut pod/seed in the class. Then fill in the equal increments between those measurements to make 14 different groups. Then take a class count of how many nut pods fit in each of those categories. Write this data in Data Table 3.
5) Measure your pinky length from the crevice next to it (without stretching the skin webbing between your pinky and your ring finger) to the tip. Do not count the fingernail. You might want to total your data for the whole day to have enough numbers to show good data. Have your teacher keep your measurement on their roster. What is the length of your pinky?

6) Find the measurement of the shortest pinky of all the classes and the longest pinky of all the classes. Then fill in the equal increments between those measurements to make up 13 different groups. Then take a class count of how many pinkies fit in each of those categories. Write this data in Data Table 4. Then fill in the pinky data and let the students copy the all-day count the next day.

7) Make a graph of the class counts of tree leaf data on Graph 1.
8) Make a graph of nut pod/seed length data on Graph 2.
9) Make a graph of the pinky data for the whole day for Graph 3.

Data Table 1

	1	2	3	4	5	6	7	8	9	10
Leaf Length (mm)										
Nut Pod length (mm)										

Data Table 2

Leaf Length (mm)													
Class Count													

Data Table 3

Nut Pod Length (mm)													
Class Count													

Data Table 4

Length of Pinky (mm)													
Class Count													
All-day Count													

Graph 1

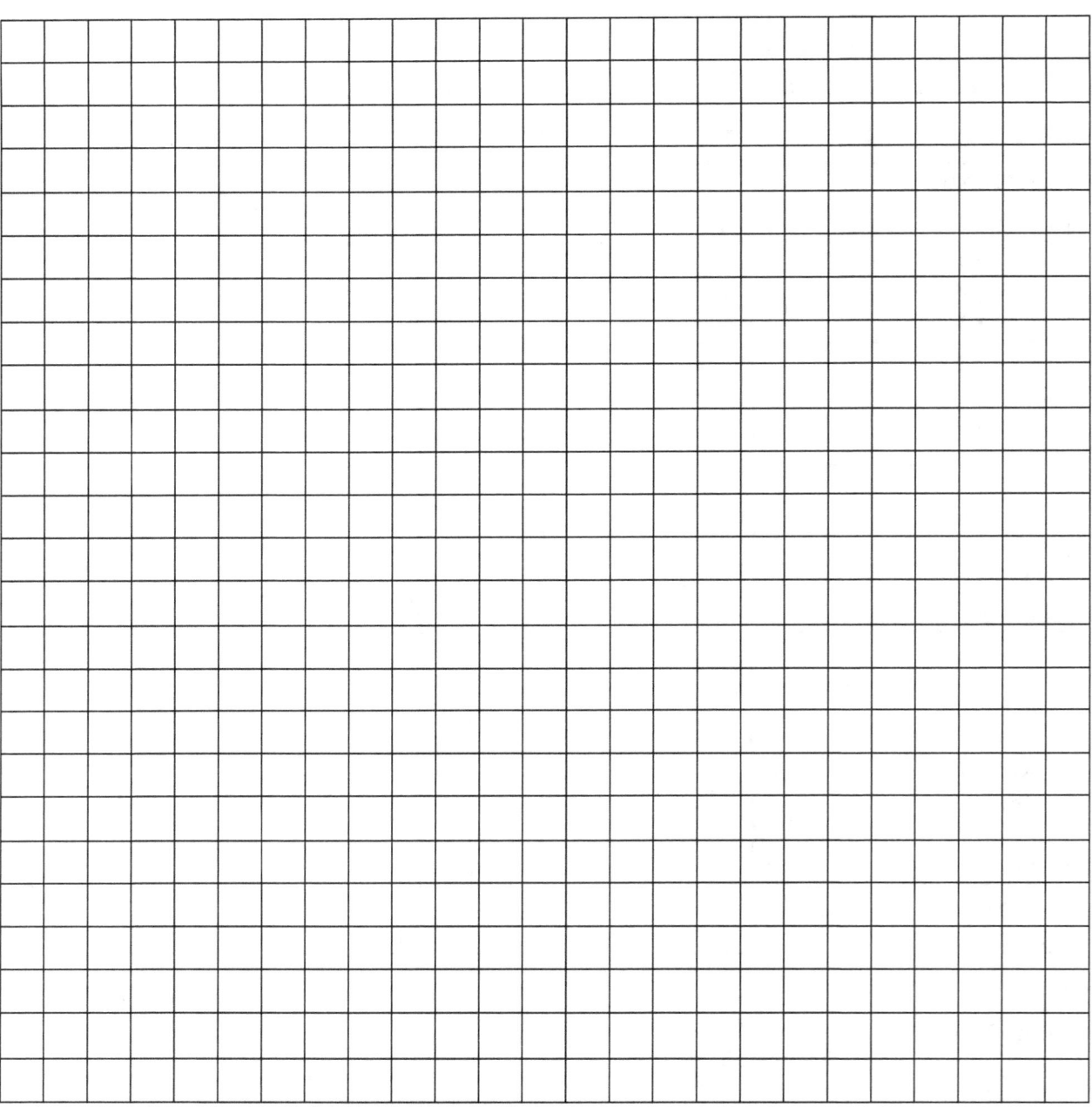

Graph 2

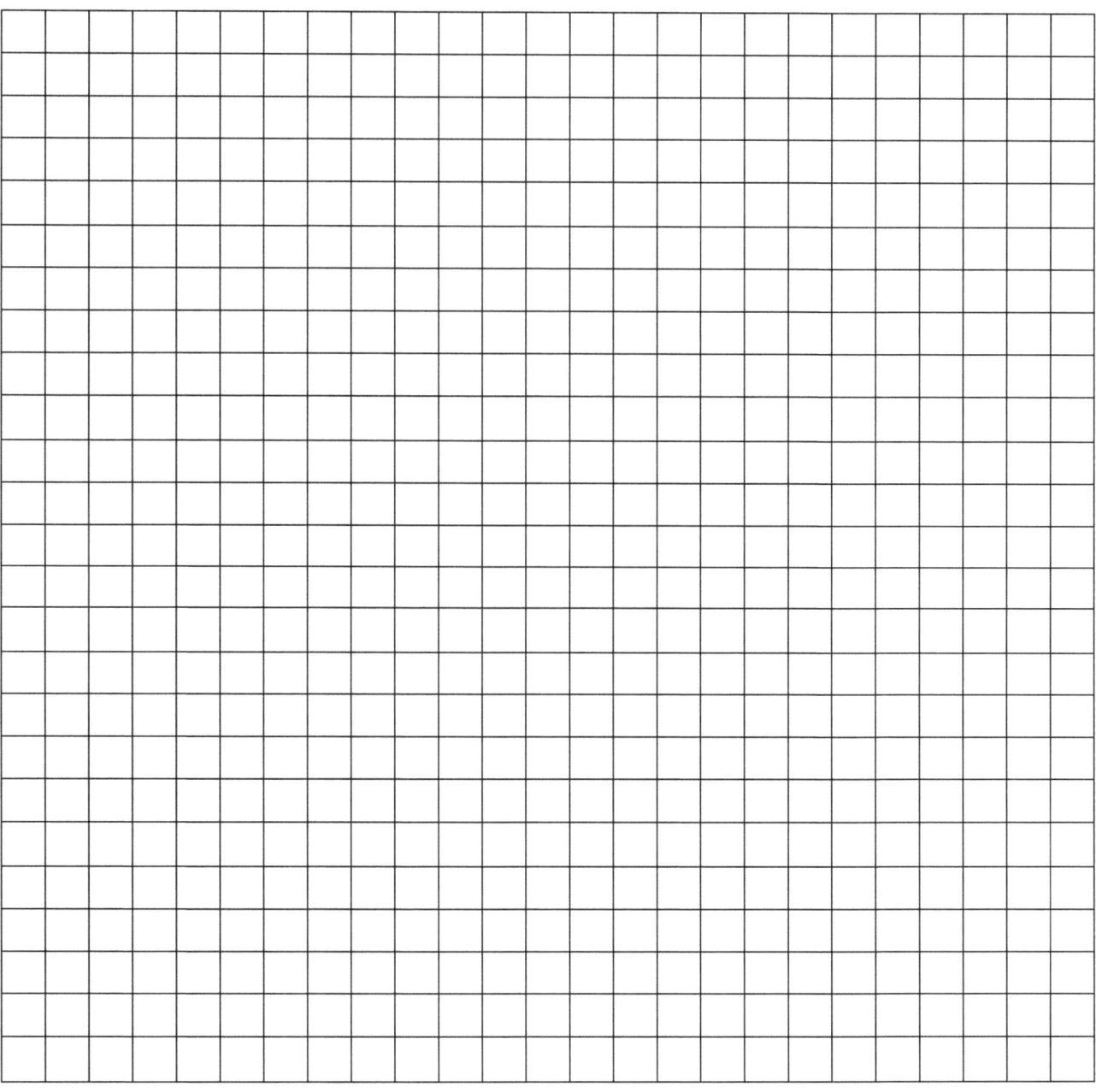

Graph 3

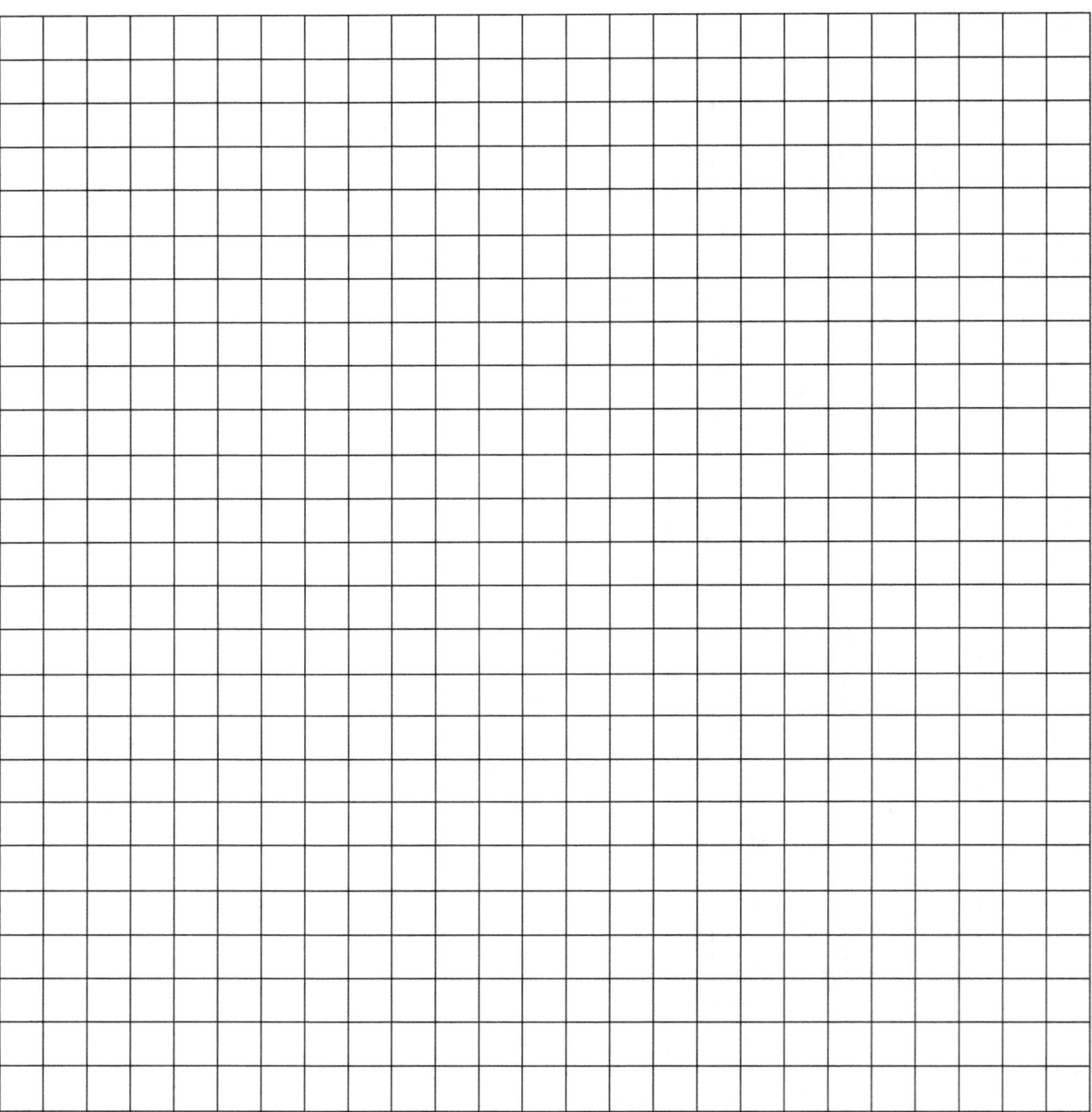

Questions:

1) What was the shape of each of those graphs?

2) Were the measurements the same for all the leaves, seeds, and pinkies?

3) Why do you think populations have variation?

4) How can this be an advantage?

5) A certain trait is selected by nature to fit best in the environment in natural selection. How can having a variety help a population survive when there is a sudden change in the environment?

6) Will the population look the same 1000 years after the environment's change selected new successful individuals? (Explain)

7) If the population changes even slightly, evolution takes place. Does evolution take place if an individual dies? (Explain)

8) Does evolution take place if an individual is born? (Explain)

9) Where does evolution happen (to the individual or a population)? Explain.

10) How could big leaves become an advantage?

11) How could big leaves become a disadvantage?

12) How could long toes become an advantage?

13) How could long toes be a disadvantage?

14) How does variation set up a population for speciation?

15) What could be some sources of error in this investigation?

Goldfish Evolution

Directions:

You will need **food serving gloves** for the teacher, a **large mixing bowl, paper plates, cheese-flavored Goldfish Crackers**, and **pretzel flavored Goldfish Crackers**. Looking at the materials and lab we will be using, what are the safety precautions we should take to protect ourselves and materials during the investigation?

In this activity, students will represent predators, a goldfish-eating shark, which selectively preys upon goldfish in small populations. This shark likes to eat two kinds of fish: **yellow fish (cheese-flavored)** and **brown fish (pretzel flavored)**. The yellow fish are easy for you to see, so they are easy to catch and eat. Brown fish travel more quickly and can evade capture more easily. Because of this, you eat only yellow fish, unless there are no yellow fish around, in which case you eat the brown fish. Fish are replaced with individuals randomly selected from an ocean (mixing bowl full of Goldfish crackers). Brown fish is determined by the presence of a dominant allele (B), and yellow fish by a recessive allele (b).

1) Send one student from your group with a paper plate to collect a <u>random</u> population of 10 fish (crackers) from the mixing bowl (ocean). Your teacher will place them on your plate for you.
2) In Data Table 1, for generation 1, record the number of yellow and brown fish in the population.
3) Choose three yellow fish from the population and eat them. If you do not have any yellow fish, fill in the missing number by eating the brown fish for a total of 3 fish eaten.
4) Send one student to the bowl (ocean) to get three more random fish and add them to your population.
5) In Data Table 1, for generation 2, record the number of yellow and brown fish.
6) Repeat steps 3-5 until you have data for all five generations.

Data Table 1

Generations	#of Gold Fish	# of Brown Fish	% of Gold Fish	% of Brown Fish
1				
2				
3				
4				
5				

7) Using the information from Data Table 1 on page 209, plot your data on the graph below to show how your population changed over time. For each generation, plot two separated bars: use one color to represent the percent population of goldfish, and use a different color to plot the percent population of brown fish.

Graph 1

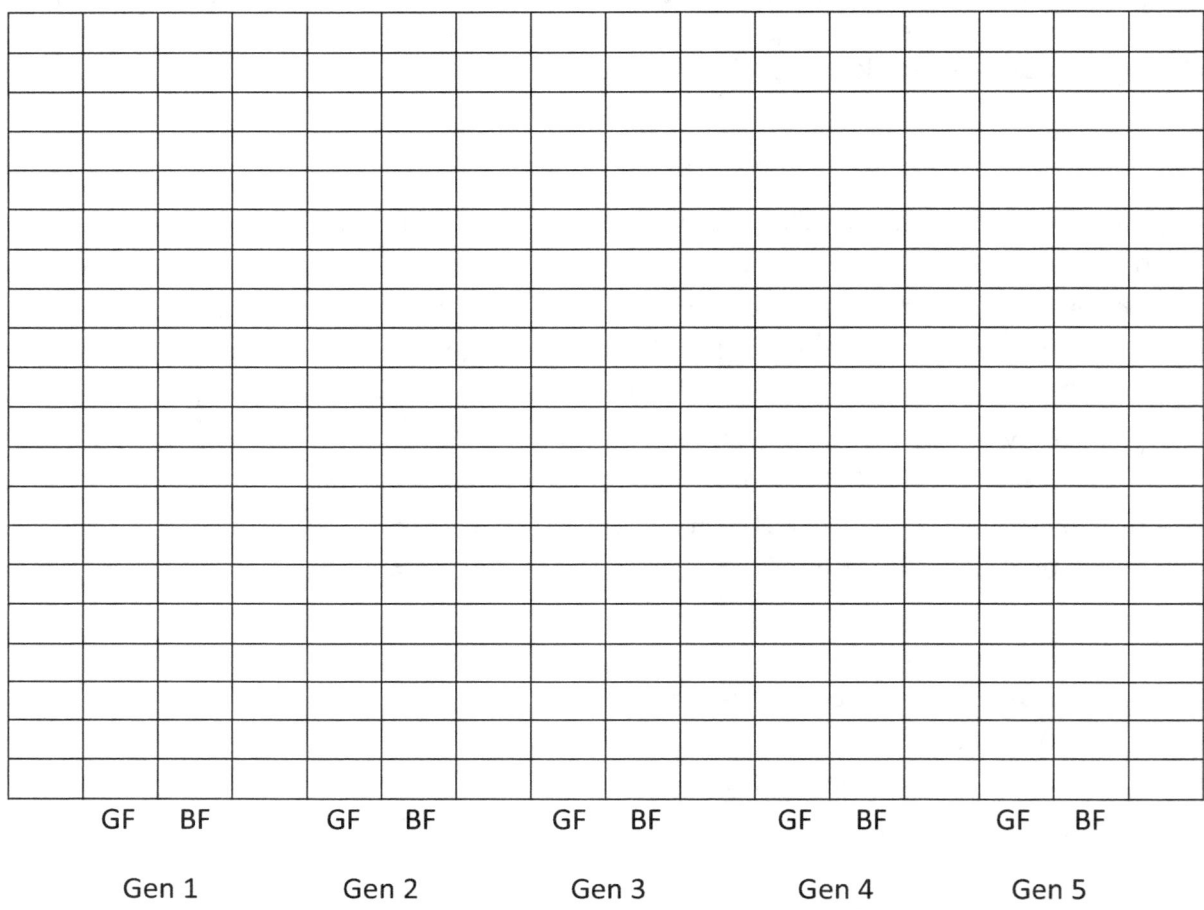

| GF BF | GF BF | GF BF | GF BF | GF BF |
| Gen 1 | Gen 2 | Gen 3 | Gen 4 | Gen 5 |

Questions:

1) How did the number of yellow fish change from generation 1 to 5?

2) Which phenotype was reduced in this population over time? Why?

3) What event occurs if there is a change in a population over time?

4) Explain what would happen over time if the brown fish were easier to catch?

5) What would happen if both fish were equally easy to catch?

6) How does this model show the Theory of Evolution and how speciation can occur?

Simple Earth & Space Science Investigations Seven Sides Publishing

Changing Environments for Beads

Directions:

You will need **red, white,** and **blue beads** in a **bowl, red, white,** and **blue construction paper,** and **colored pencils. Looking at the materials and lab we will be using, what are the safety precautions we should take to protect ourselves and materials during the investigation?**

1) Have each group randomly get 10 beads out of the bowl and place them on white paper. The beads will represent a population with three variants in them, and the paper will represent the environment they are in. Count how many beads there are of each color and write this in Data Table 1 for the first generation.
2) The students will represent a predator of the beads. Have the students in each group take out three beads that do not match the environment's background, place them back into the bowl, and randomly pick three more beads out of the bowl (if you do not have any that don't match the background take ones that do match until you have three).
3) Add the three new beads to the paper, count how many beads there are for each color in the population, and write this down in Data Table 1 for Generation 2.
4) Repeat steps 2 and 3 for six more generations.
5) After completing eight generations change the white background to red or blue and predict how your population will change over time.
6) Repeat steps 2 and 3 for eight generations and write this data in Data Table 2.
7) Once you have completed both Data Tables, graph your data for Data Table 1 on Graph 1 and Data Table 2 on Graph 2, making a line graph using red (for red beads), black (for white beads), and blue (for blue beads) colored pencils.
8) Once the graphs are completed, answer the questions that follow.

Data Table 1

Bead Color	Gen 1	Gen 2	Gens 3	Gen 4	Gen 5	Gen 6	Gen 7	Gen 8
Red								
White								
Blue								

Data Table 2

Bead Color	Gen 1	Gen 2	Gen 3	Gen 4	Gen 5	Gen 6	Gen 7	Gen 8
Red								
White								
Blue								

Graph 1 (Line Graph) Color of Background: White

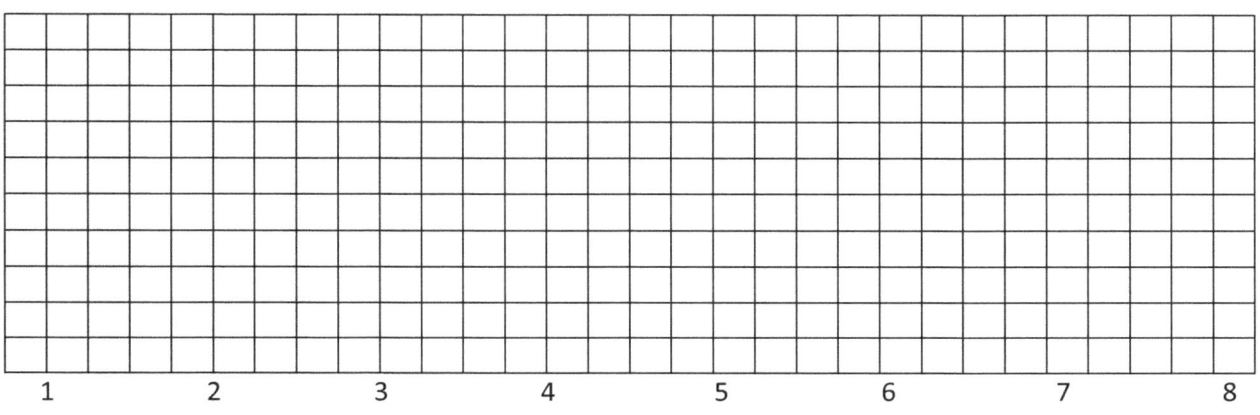

Number of Generations

Graph 2 (Line Graph) Color of Background: _____

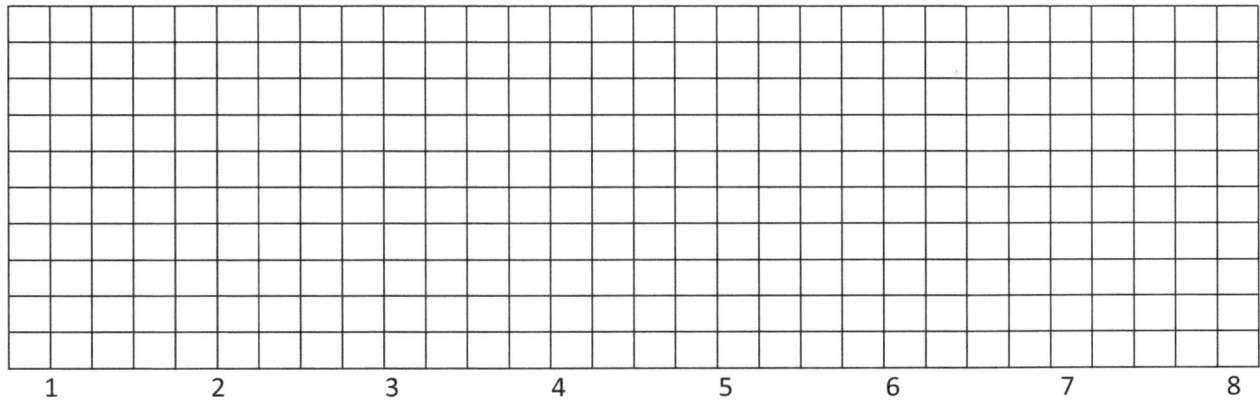

Number of Generations

Questions:

1) How did the population of beads change when there was a white background?

2) Why would this be useful in nature?

3) How did the population of beads change when there was a red or blue background?

4) Why would variations in a population benefit a population when an environment changes?

5) How could having variations in a population be hazardous to some individuals in the population?

6) What could happen to this population if the background turned green?

7) Which population would be more fit, one with little variation or one with lots of variation? Explain why.

8) How was this a good model for showing how variations within the population help populations adapt to changing environments?

9) How was this model not accurate?

Causes of Invasive Species

Directions:

Use the **internet** and your **textbook** to research invasive species and answer the following questions.

1) Give three examples of how climate change is causing invasive species.
    a.

    b.

    c.

2) Give three examples of how humans introduce invasive species.
    a.

    b.

    c.

3) Give three examples of when humans are the invasive species.
    a.

    b.

    c.

Humans Changing Ecosystems

Directions:

Use the **internet** to research how ecosystems changed when humans added or removed organisms from an ecosystem. Include which organisms were affected and how the food webs changed.

1) Humans moved to an uninhabited island, Mauritius, east of Madagascar, and brought pets.

2) Humans move into a new area and clear the land to build houses.

3) Humans release Burmese pythons in the Florida Everglades.

4) Humans release Asian carp into the Mississippi river to eat plants covering the river's bottom.

5) Humans plant crops on unused land.

6) Nutria were brought from South America to North America for the fur trade.

7) Wild boars were brought to the US for hunting and food.

8) Domesticated cats were bred and then released into the environment.

9) Humans chop down trees in the rainforest to make a field to raise cattle in Brazil.

10) In 1890 and 1891, European starlings were introduced to Central Park in New York City.

What Happens to the Food Web?

Directions:

Use the food web below to predict what would happen to the populations of the organisms when a change happens to the ecosystem.

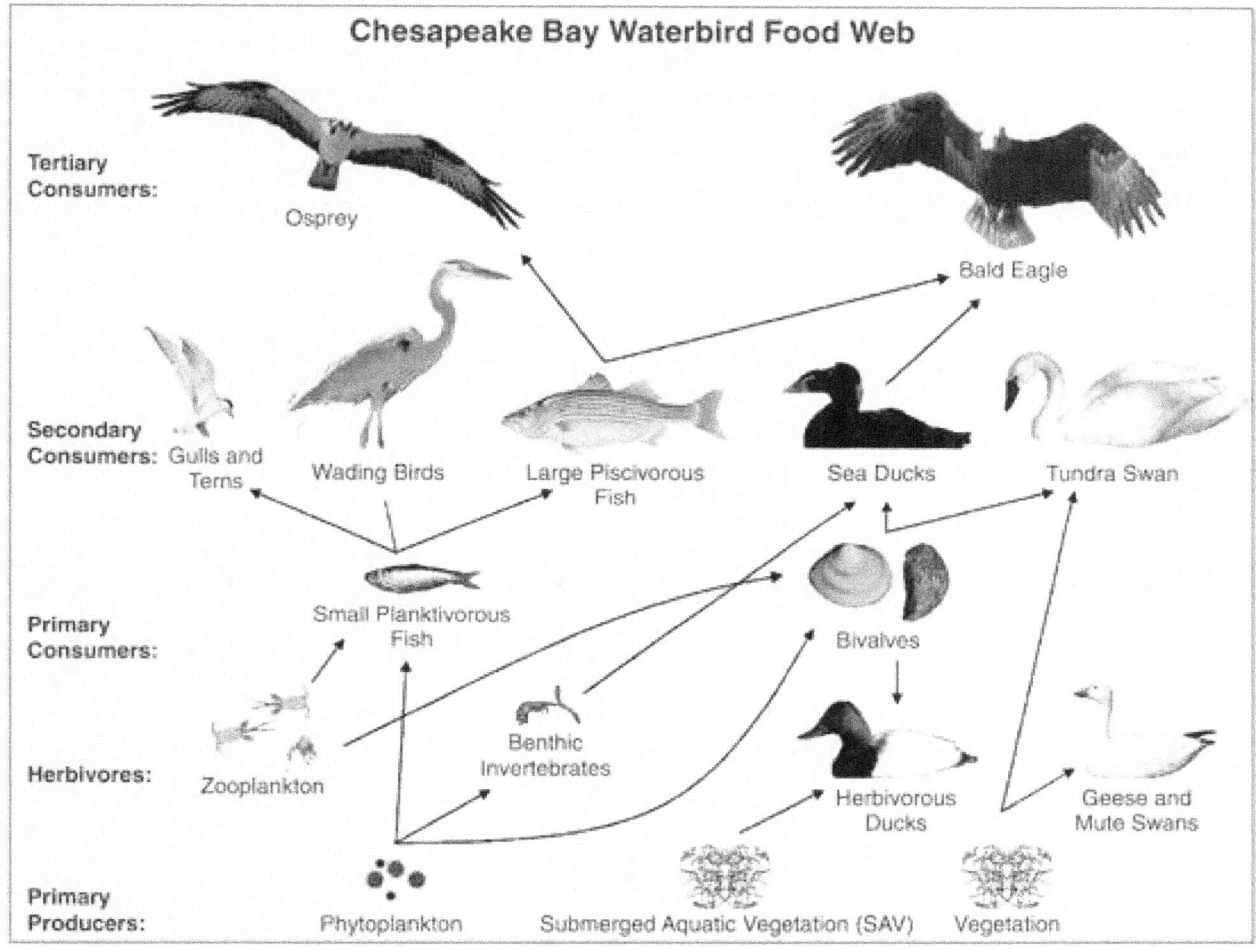

Picture from http://commons.wikimedia.org

Questions:

1) What would happen to the Zooplankton population if the Bivalves were to go extinct?

2) What would happen to the fish populations if wading birds and gulls left the area?

3) What would happen to the populations of land animals if domesticated dogs were left in the area to breed?

4) What would happen to the populations if chemicals were released into the water that killed all the phytoplankton?

5) This community is in a cold coastal marine ecosystem. What would happen to each population if climate change made it warmer and brought more fish to the area?

6) What would happen to the bird populations if the fishing industry were to overfish the area?

   a. How would that affect the zooplankton population?

   b. How would that affect the osprey population?

7) What would happen to the geese and swan populations if Asian carp were introduced and adapted to this community and ate all the plants?

   a. What would that do to the bivalve populations?

Our Little Mountain

Before I was born, my parents used to go on hikes and have picnics on this small mountain that overlooks a pass between two large mountain ranges. It was filled with tall pine trees with a clearing on the top. One day, shortly after I was born, we drove out to their special spot and saw that bulldozers were getting ready to knock down and uproot the trees. My parents were upset to see this. My dad saw a trailer parked close by with a makeshift parking lot. He went inside to find out what was going on. When he came out, he had a piece of paper in his hand and a scary smile on his face. He told my mom that they were selling land plots and were going to be building a winding road and houses on this mountain. My dad had just purchased the 5-acre lot at the top, which was their special picnic spot. We lived in Golden, Colorado, and he thought it could be a great place to put a vacation home if they could ever afford it one day.

A month later, we came back to see how the construction progressed. All the trees and bulldozers were gone, and a gravel road wound back and forth up our little mountain. My parents had packed a picnic that we would eat on our land at the top. When we had our picnic, the wind kept blowing dirt up into the air, and clouds of dust kept getting in our food and eyes; it was not very enjoyable.

Three years later, after my sister was born and able to walk, we went back to look at our mountain again. There was the greenest grass you ever saw up and down the mountain. Two houses were built, one at the bottom and an A-frame part way up. The one at the bottom served as an office for selling the plots on the mountain. We drove up the mountain and had a great time running around in the grass on our 5-acres. A crow flew down and snatched the peanut butter sandwich right out of my sister's hand. Everyone laughed except for her. Later we saw some deer walk out of the woods next to the land cleared on the mountain. I looked out and imagined John Wayne driving thousands of head of cattle through the pass our land overlooked. My dad got a promotion a few weeks later, and we moved to Wisconsin for his new job.

Ten years later, my dad thought it would be fun for us to go on a family vacation, as he saw in a movie. He wanted to wind around America and eventually see our mountain. We noticed lots of bushes and small trees were all over the mountain when we got there. The house at the bottom was abandoned. My dad did some research while we were there and found out we were only one of two that had purchased any plots, and the developer went bankrupt, and the mountain was abandoned. We went to the top of the mountain and picked berries growing on the bushes there. We saw a few rabbits and a small red fox running down the mountain across the dirt road just before leaving. I thought this could be a great place to put a house one day.

Twenty years later, I was now married with three children. My wife and I thought it would be fun to take our family skiing. I told her that my family owned some land on the top of a little mountain not far away from the lodge where we were staying. While on vacation, one afternoon, we took a drive to find the mountain. We found the little dirt road off the highway. As we drove up, we saw many leafy trees filling the mountain. We drove up the mountain until we reached the end of the road at the top. We walked through the forest until we reached a clearing at the top of the mountain, looking over the large valley pass that separated the mountain ranges on either side. I remembered the other times I had been there when I was younger. I wondered why my parents never built a house up there. There were lots of squirrels and birds. Pine tree seedlings seemed to be taking hold and growing around the clearing.

When I had retired, and our oldest was visiting with our grandson, she asked about the land we visited in Colorado. My parents had died a few years before when COVID 19 hit, and I decided to look through the lockbox my father had given me before their passing. I noticed the deed on the land had been amended. The other owner moved off the mountain when the trees blocked their view of the valley below. My father was sold all the land surrounding the mountain. I now owned our little mountain. My publishing business took off, and I had enough money to build a house on the top of that clearing on the mountain. I hopped on a plane, went straight to our mountain, and saw that most of the leafy trees were gone. There were lots of small and medium-sized pine trees covering the mountain. I hired a construction company and built a large two-story cabin with big picture windows looking over both sides of the mountain. Because it was so far away from the nearest town and it was on the top of the mountain, it took a few years to build. My wife and I moved there after our youngest son graduated college. Our house also became the rest of our family's vacation home my parents dreamed about when I was just a baby.

I was 95 years old when I died. My last view of this Earth was the tall pine trees that were just like the ones I saw when I was a baby. The trees filled the whole mountain like they did when my parents went on their hikes and picnics on the top of the mountain. These trees were the bottom of the frame of our bay window in the front of our house. I do not know if I imagined it or not. But I thought I saw a herd of cattle going through the pass in the valley between the two mountain ranges just before I died.

Directions: Read each section of the story and draw a picture of the story's description in each box. When done with the story and your drawings, you should have a good picture of how succession looks.	Before I was Born
A Month Later	Three Years Later
Ten Years Later	Twenty Years Later
When I Retired	I was 95 Years Old

Questions:

1) What happened to the biodiversity when the land developers knocked down the forest?

2) During the story, when did the mountain have its greatest diversity? Explain why.

3) What happened to the plant populations when the forest was knocked down?

   a. Were there any homes for the animals there?

   b. What happened to the animal populations?

4) When did the grass populations increase?

5) When the trees started growing, what happened to the population of the grasses?

6) What happened to the kinds of plants during the story?

7) What happened to the kinds of animals during the story?

8) Who was the invasive species in the story, and how did they affect the ecosystem populations when they arrived?

9) From this story, can life find a way to recover from change? Explain your answer.

10) Who do you think will be outlived, humans or the rest of life? Explain.

Primary or Secondary Succession?

Directions:

Go outside, and find five examples of each.

1) Primary Succession takes place where rock is taking in life and eroding it away to make soil.

   a.

   b.

   c.

   d.

   e.

2) Secondary Succession occurs when something interrupts the ecosystem, reestablishing itself and keeping the soil.

   a.

   b.

   c.

   d.

   e.

Virtual Investigations that go with Ecological Changes

ExploreLearning.com

Food Chain

Prairie Ecosystem Gizmo

Forest Ecosystem Gizmo

Rabbit Population by Season Gizmo

Ecosystems STEM Case

Ecosystems Handbook

Coral Reefs 1 – Abiotic Factors

Coral Reefs 2 – Biotic Factors Gizmo

Natural Selection

Evolution: Mutation and Selection

Evolution: Natural and Artificial Selection

Rainfall and Bird Beaks

Evolution STEM Case

Evolution Handbook

Human Evolution – Skull Analysis

Microevolution

Cladograms

Effect of Environment on New Life Form Gizmo

Estimating Population Size Gizmo

Phet.colorado.net

Natural Selection

Unit 8: Managing Natural Resources

Renewable Resource
- solar → ecosystem
- wind → ecosystem
- biomass → air, soil, ecosystem
- hydroelectric → surface water, ground water
- geothermal → Renewable Resource
- natural gas → Renewable Resource

Non-renewable Resources
- nuclear → Non-renewable Resources
- petroleum → fossil fuel
- coal → fossil fuel
- fossil fuel → air, soil, Non-renewable Resources

Soil Moisture

Directions and Questions:

You will need a **soil sample** that fills a **plastic shoebox** and a **soil moisture probe** attached to an **interface** connected to a **computer** with **Logger Pro**. **Looking at the materials and lab we will be using, what are the safety precautions we should take to protect ourselves and materials during the investigation?**

1) Hold the soil moisture probe so that the two forks are positioned vertically with each other. Gently push them into the soil, first down, then horizontally, until the entire probe is under the soil. The probe will measure how much water it contacts. What is the soil moisture value of your soil sample?

2) While watching the value displayed on the computer, squeeze the sides of the box; what happened to the soil moisture value when you squeezed the box?

    a. Why do you think this happened?

3) Why is soil moisture important?

4) Which type of soil do you think is better at holding this moisture?

5) Which type of soil do you think is better for draining this moisture?

6) Which type of soil does not like absorbing moisture and keeps it above it helping make ponds and lakes?

Infiltration Rate & Water Holding Capacity

Directions:

You will need one **small beaker** (100-150 mL to measure soil with), two **larger beakers** (400-600 mL), a **graduated cylinder**, a **soil sample**, a **sand sample**, four **coffee filters, rubber bands**, and a **scale. Looking at the materials and lab we will be using, what are the safety precautions we should take to protect ourselves and materials during the investigation?**

1) Take two large coffee filters and place them together in a beaker about half full of water. Remove the filters and gently wring out any excess water. Record the mass of the damp coffee filters in Data Table 1. Repeat this process for the second set of two coffee filters. (Note: we need to use two coffee filters stacked together so that they do not break during the lab. If you use one, it will break during the lab.)
2) Use the 100 mL beaker to measure approximately 40 mL of soil to the damp filter. Measure the mass of the damp filter and dry soil sample; record this in Data Table 1. Repeat this process for the sand sample.
3) Get the two larger beakers (400-600 mL) and label the first soil and the second sand.
4) Carefully place the filter and soil sample into the beaker labeled soil about halfway into the beaker and hold the filter in place with a rubber band. Repeat this procedure for the sand sample.
5) Use the graduated cylinder to measure and slowly add 25 mL of water to the soil sample. Make sure you add it slowly, so the water goes down through the soil and not the filter's sides. Repeat this procedure for the sand sample.
6) After five minutes, carefully remove the coffee filters and soil (keep the soil contained in the filters) and measure the amount of water left in the beaker. Record this information in Data Table 1. Repeat this procedure for the sand sample.
7) Then add 25 mL of water into each beaker. Place the coffee filter with the soil contained in it into the bottom of the beaker. Repeat this procedure for the sand and let them sit in the bottom of the beaker until the soil has become saturated.
8) Carefully remove the beaker's coffee filters and soil (keeping the soil in the coffee filter). Let any excess water drain from the soil by holding it over the beaker. When the soil has stopped draining, measure the mass of the saturated soil and filters. Record this in Data Table 1. Repeat this process for the sand sample.

9) Subtract the damp filters' mass from the saturated samples and filters to find the saturated soil's mass.
10) Subtract the dry sample mass from the saturated sample to find the mass of water in the saturated sample.
11) Determine the percent water holding capacity by taking the mass of water in soil or sand divided by the mass of the saturated soil or sand multiplied by 100.

Data Table 1

	Soil Sample	Sand Sample
Mass of damp filters		
Mass of damp filters and sample		
Mass of dry sample		
Water added	25 mL	25mL
Water left in the beaker, infiltration rate		
Mass of Saturated sample and filter		
Mass of saturated sample		
Mass of water in the saturated sample		
Water holding capacity %		

Questions:

1) Which sample drained more water?

2) Describe the difference between the infiltration rate of soils with high clay composition and those with high sand composition.

3) Describe the difference between the water holding capacity of soils with high clay composition and high sand composition.

4) Which sample would you want on a golf course to get rid of rainwater fast so that you don't lose money by having golfers have to stay off the golf course after getting some rain?

5) What could be some sources of error in this investigation?

6) What careers would be interested in soil types, infiltration rate, and water holding capacity?

Managing Garden Soil Moisture

Directions:

You will need a **heat lamp**, two shoebox-sized **tubs** of **soil** from the same source, one with some **mulch** covering the top and one without mulch. You will also need a **soil moisture probe** attached to an **interface** connected to a **computer** with **Logger Pro. Looking at the materials and lab we will be using, what are the safety precautions we should take to protect ourselves and materials during the investigation?**

1) Pour 500 mL of water evenly into each of the boxes of soil. Let it sit for a few minutes and measure the amount of soil moisture in each box by placing the soil moisture probe into the soil where the two prongs are vertical to each other and carefully pushing the probe into the soil until it is entirely under the soil. Write this data in Data Table 1.
2) Let the soil samples sit under a heat lamp for three-four days.
3) Then measure the moisture using the soil moisture probe again for each box of soil. Write this data in Data Table 1.

Data Table 1

	Soil With Mulch	Soil Without Mulch
Moisture Before		
Moisture After		
Change in Moisture		

Questions:

1) Which sample of soil held more moisture in it?

2) Which sample would you want to be set up for your bushes and flowers at your house? Explain Why.

3) What is the main purpose of mulch?

4) Which setup would allow you to use less water to keep your plants alive?

5) What are three materials in our area that are used for mulch?

6) Where can you get sources for those materials that you can use as mulch?

7) What type of mulch will you put on your plants in the future? Explain why.

Organic Gardening and Hydroponics

Directions:

Use the **internet** to research Organic Gardening and Hydroponics to help you answer the questions that follow.

1) What is organic gardening?

2) What are some examples of organic gardening?

3) What are the benefits of organic gardening?

4) What are the negative aspects of organic gardening?

5) What is hydroponics?

6) What are some examples of hydroponics?

7) What are the benefits of hydroponics?

8) What are the negative aspects of hydroponics?

Sustainability of Diets for a Growing Population

Directions:

Use resources provided by your teacher and the **internet** to help you answer the following questions.

1) What is happening to the world's population numbers?

2) Which trophic level in an ecosystem has the most energy available to it, and which has the least?

3) What is the energy/feed input that we put into livestock?

4) Where can humans get protein?

5) What does the literature say are the most efficient (spend the least amount of money and resources to get the most) protein sources for humans to eat?

6) Why are they considered efficient?

7) What does the literature say are the least efficient sources of protein for humans to eat?

8) Why is it considered inefficient?

9) China and India are the most densely populated countries in the world. They already know and understand this topic because they need to be able to feed their people. Where do they get the majority of their protein, and how does it affect their culture?

10) Fifty years from now, if you were put in charge to help feed an overly populated country and get them enough protein, describe the type of diet you would try and instill in the people that would be sustainable.

11) What careers would help deal with this issue?

Nonrenewable Resources Chart

Directions:

Use the **internet** and your **textbook** to research and fill out this chart on nonrenewable energy resources.

Type	How we obtain and transport it	How we use it	How it affects the environment	How did it form?
Petroleum				
Coal				
Natural Gas				
Nuclear				

Questions:

1) What are the advantages of using petroleum?

2) What are the disadvantages of using petroleum?

3) What type of reaction goes into burning petroleum products for energy?

    a. Give an example of a reaction using a petroleum product for energy.

    b. Are there any products from this reaction that are harmful to the environment? If so, how are they harmful?

4) What are the advantages of using coal?

5) What are the disadvantages of using coal?

6) What type of reaction goes into burning coal for energy?

    a. Are there any products from this reaction that are harmful to the environment? If so, how are they harmful?

7) What are the advantages of using natural gas?

8) What are the disadvantages of using natural gas?

9) What type of reaction goes into burning natural gas?

   a. Are there any of the products from this reaction that are harmful to the environment? If so, how are they harmful?

10) What are the advantages of using nuclear energy?

11) What are the disadvantages of using nuclear energy?

12) Are there any products from nuclear reactions that are harmful to the environment? If so, how are they harmful?

13) Describe some careers involved with the exploration, extraction, production, and disposal of these resources?

Renewable Resources Chart

Directions:

Use the **internet** and your **textbook** to research and fill out this chart on renewable energy resources.

Type	How we obtain and transport it	How we use it	How it affects the environment
Wind			
Solar			
Hydroelectric			
Geothermal			

Questions:

1) What are the advantages of using wind energy?

2) What are the disadvantages of using wind energy?

3) What are the advantages of using solar energy?

4) What are the disadvantages of using solar energy?

5) What are the advantages of using hydroelectric energy?

6) What are the disadvantages of using hydroelectric energy?

7) What are the advantages of using geothermal energy?

8) What are the disadvantages of using geothermal energy?

9) Describe some careers involved with the production of energy using these resources.

Nuclear Reactor

Directions and Questions:

Use the **internet** and your **textbook** to research how nuclear reactors work.

1) Draw a diagram of the main parts of a nuclear reactor showing how it works.

2) Where does the reaction take place in the reactor?

3) How is the water there used?

4) How does this make electricity?

5) How do we cool off the reactor so it does not explode?

6) How do we slow the reaction down so it does not explode?

7) What is the fallout from a nuclear reactor meltdown?

8) What do we do with the waste products?

9) Why does the US prohibit the production of any new nuclear power plants?

10) How do other countries dispose of their nuclear waste?

11) What are the positive and negative effects of using nuclear power plants to produce electricity?

12) Why do we not want developing countries to develop this technology?

Natural and Manmade Disasters

Directions:
Use the **internet** to fill in the following charts while researching natural and manmade disasters.

Natural Disasters

Natural Disasters	What it Does	Environmental Impacts	Disasters that cause it	Disasters it can cause
Hurricane				
Flood				
Tornado				
Earthquake				
Volcanic Eruption				
Tsunami				
Mudslide Avalanche				
Drought				
Wildfire				
Deforestation				

Manmade Disasters

Manmade Disasters	What it Does	Environmental Impacts	Disasters that cause it	Disasters it can cause
Oil Spill				
War				
Plane Crash				
Nuclear Meltdown				
Nuclear Bomb				
Mining Accident				
Municipal Development				
Fire				
Global Warming				
Fracking				

Worldwide Disaster Presentations

Directions:

1) Use the **internet** and your **textbook** to create two presentations in a digital format designated by your teacher (examples: Animoto, Prezi, PowerPoint, or Word). Make one for a Natural disaster and one for a Manmade Disaster.

2) Follow the rubric below and include pictures, videos, information from your charts, and any other information you think is important or interesting.

Poster Title: _____

Rubric

Scores range from 0-10 in each box.

Title: Include the name of the disaster and the author's name 0, 5, or 10	Accuracy of the information 0-10	What happens in the disaster 0-10	What other disasters can cause it 0-10
What other disasters it can cause 0-10	Environmental Impacts of the disaster 0-10	Impact of disaster on humans 0-10	Short term effects 0-10
Long term effects 0-10	The overall feeling of the presentation and its appearance 0-10	Total Points 100 points possible	

Student's Name: _____

How to Dispose of Waste

Directions:

Use the **internet** to research what happens to your garbage and sewage after leaving your house. What are some careers involved with developing this process? Present this on any electronic/digital media your teacher chooses.

> 100 points
> <u>Rubric:</u>
>> 30 points: for mentioning the different types of materials in the garbage and waste that leaves the house
>> 30 points: for tracing what happens to each type of material after it leaves your house
>> 30 points: include garbage and sewage
>> 10 Points: for completeness and accuracy of information

Possible Waste Management Resources

Waste Management and Energy

- http://www.focusforwardfilms.com/films/11/the-landfill

Waste Management Learning Center Website

- http://www.wm.com/thinkgreen/learning-center.jsp

Anatomy of a Landfill

- http://www.wm.com/thinkgreen/pdfs/Anatomy_of_a_Landfill.pdf

Bioreactor Landfill

- http://www.wm.com/thinkgreen/pdfs/bioreactorbrochure.pdf

Waste Management Regulations

Directions:

Use the **internet** to research regulations on Landfills, burning, and recycling. Include when and where the landfills are used and what is allowed to go into landfills. Include when and where burning is allowed and what is allowed to be burned. Include what is allowed to be recycled and how those materials are recycled and then can be used. What are the careers involved in developing these processes? Present this on any electronic/digital media your teacher chooses.

    100points
    Rubric:
- 10 points: for where landfills are and are not allowed
- 10 points: for what materials are and are not allowed in the landfills
- 10 points: for when burning is and is not allowed
- 10 points: for where burning is and is not allowed
- 10 points: for what is and is not allowed to be burned
- 10 points: for what is allowed to be recycled
- 10 points: for how those materials are recycled
- 10 points: for what the recycled materials will be used for
- 10 points: for accuracy and completeness of the information
- 10 points: for overall neatness and feel of the presentation

Possible Waste Management Resources

Waste Management and Energy

- http://www.focusforwardfilms.com/films/11/the-landfill

Waste Management Learning Center Website

- http://www.wm.com/thinkgreen/learning-center.jsp

Anatomy of a Landfill

- http://www.wm.com/thinkgreen/pdfs/Anatomy_of_a_Landfill.pdf

Bioreactor Landfill

- http://www.wm.com/thinkgreen/pdfs/bioreactorbrochure.pdf

Waste Management Plan

Directions:

Construct a new waste management plan for your neighborhood. Ensure you include a plan for disposing of materials in garbage and sewage; and how you could recycle material. What kind of an impact will your plan have on the environment? Can any part of this plan be used to generate revenue for your neighborhood? What may be the cost for your plan to your neighborhood? Present this on any electronic/digital media your teacher chooses.

100 points

Rubric:

- 10 points for how you would dispose of materials in the garbage
- 10 points for how to dispose of sewage
- 10 points for how you would recycle each material to be recycled
- 10 points for how your recycled material will be used
- 10 points for how your plan would realistically impact the environment
- 10 points for how much your plan would cost the neighborhood
- 10 points for generating revenue for your neighborhood
- 10 points for neatness
- 10 points for completion and accuracy
- 10 points for an overall feel

Possible Waste Management Resources

Waste Management and Energy

- http://www.focusforwardfilms.com/films/11/the-landfill

Waste Management Learning Center Website

- http://www.wm.com/thinkgreen/learning-center.jsp

Anatomy of a Landfill

- http://www.wm.com/thinkgreen/pdfs/Anatomy_of_a_Landfill.pdf

Bioreactor Landfill

- http://www.wm.com/thinkgreen/pdfs/bioreactorbrochure.pdf

Local Clean Up

Directions and Questions:

Identify a place near you that needs to be cleaned up, gather materials, and a group together to clean this area. Fill out this sheet below to help and document your efforts.

1) Where is the area you want to clean?

2) What needs to be cleaned?

3) Is there anything that could be hazardous to those cleaning?

   a) If so, what are they?

4) Could any agency help you? If so, who?

   a) Did any agency help you?

5) What tools and materials do we need to get started?

6) Where are we going to dispose of the waste?

7) Where was the waste disposed of?

8) How many people showed up to help?

9) How did your actions change the area?

10) How do you feel after cleaning the area?

11) How does helping clean an area make you think more about how you casually dispose of waste?

   a) Will you make any new choices? If so, what are they?

Virtual Investigations that go with Managing Natural Resources

ExploreLearning.com:

    Energy Conversions

    Household Energy Usage

    Water Cycle

    Carbon Cycle

    Nitrogen Cycle STEM Case

    Nitrogen Cycle Handbook

    Porosity

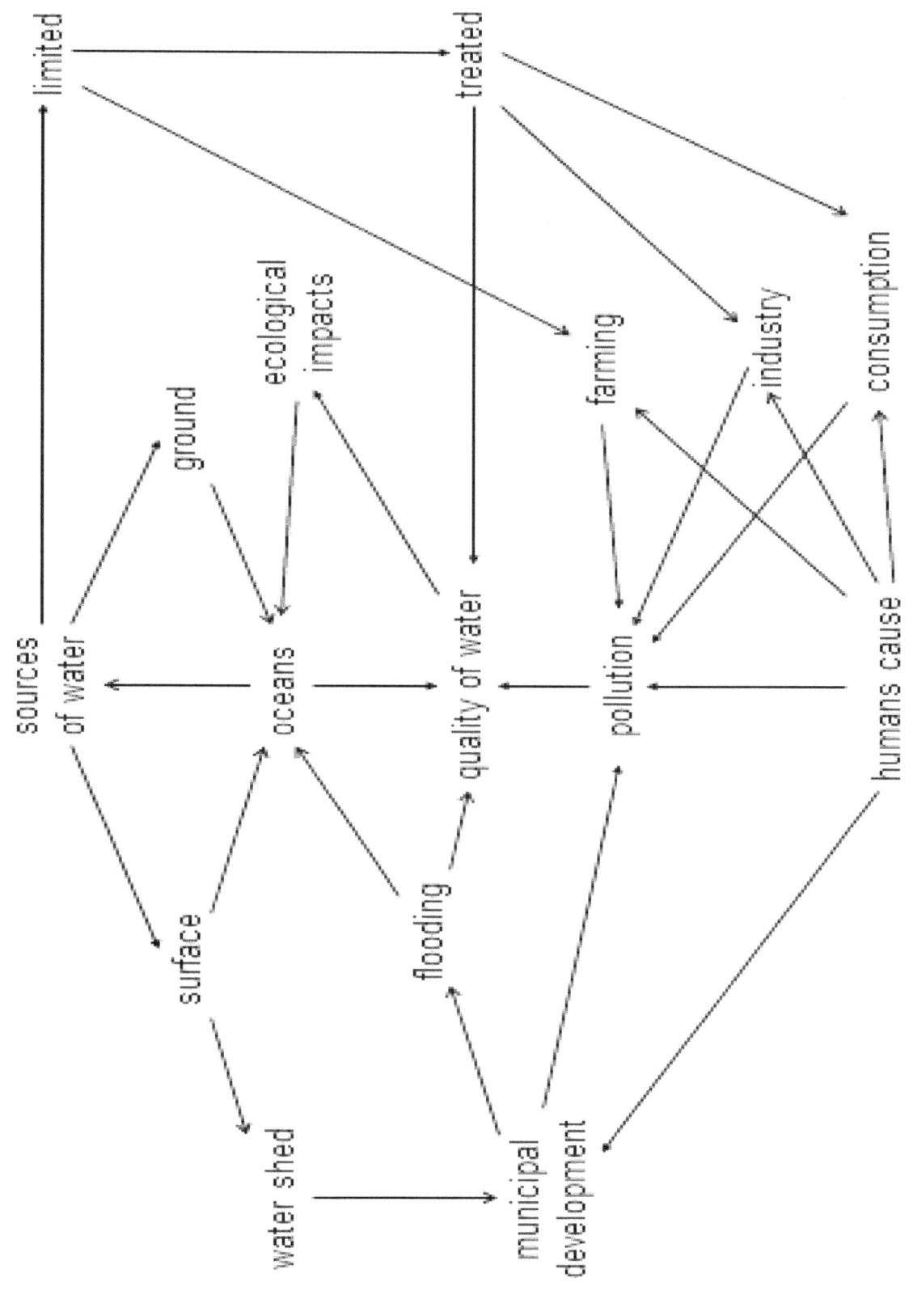

The Water Cycle (B)

Directions:

Use the **internet** or your **textbook** to draw the water cycle below. Include condensation, precipitation, runoff, groundwater storage, groundwater discharge, groundwater recharge, evaporation, and transpiration.

Questions:

1) Which type of water stored on Earth is considered a renewable resource? Explain why.

2) Which type of water stored on Earth is considered a nonrenewable resource? Explain why.

3) How does groundwater discharge and recharge affect subsidence?

Groundwater Pollution Lab

Directions:

You will need a **slice of white bread**, **red food coloring**, a **paper towel**, and a **beaker of water** with a **pipette. Looking at the materials and lab we will be using, what are the safety precautions we should take to protect ourselves and materials during the investigation?**

1) Take your slice of bread and hold it bottom up over a paper towel. Take your red food coloring and place one to two drops on the middle of the bottom crust of your slice of bread.
2) Then fill your pipette with water and slowly drop water on the spot where you just put the food coloring drop by drop until you have put two pipettes on the bread keeping the bread held on its edge for at least five minutes.

Questions:

1) Draw the pattern of the die you see on the bread.

2) Which way did the food coloring spread first?

3) As you go down the bread, did the intensity of the food coloring stay the same, get lighter in color, or get darker in color?

4) Where is the color pattern the widest at the top, middle, or bottom?

5) What does the bread represent?

6) What does the food coloring represent?

7) What do the water drops represent?

8) Did the pollutant continue through the bread, or was it filtered out?

9) Could a surface pollutant make it into groundwater? What could this mean for the person next door using well water?

10) Did the water cause the pollutant to spread out or go straight down?

   a. How does this make it difficult to locate where the pollutant originated?

11) What human activities on the surface affect groundwater quality?

12) What kinds of pollutants might come from common household products?

13) What kind of pollutants might come from septic tank fields?

14) What kind of pollutants comes from landfills?

15) Where should we not put landfills?

16) What are some other sources of groundwater pollution?

17) What careers do you think would be interested in how groundwater can become polluted?

Parts Per Million

Directions:

You will need a **well-plate**, **white paper**, two **pipettes**, two **beakers of water**, and **red food coloring**. Looking at the materials and lab we will be using, what are the safety precautions we should take to protect ourselves and materials during the investigation?

1) Place the well-plate on top of the white piece of paper. Add ten drops of red food coloring in well number 1 with a pipette. The concentration is 10/10, 1, or 100%. Fill in Data Table 1 for well one.
2) Using the same pipette, carefully take the food coloring out of the first well and place one drop in the second well. Rinse that pipette in a beaker of water and change the water in that beaker. Use the other pipette and the other beaker of water to add nine drops of water to the second well. This process makes the concentration in that well 1/10, .1, or 10%. Fill in Data Table 1 for well two.
3) Using the first pipette take one drop out of the second well and place it in well three. Use the second pipette and add nine drops of water into well three. The concentration in well 3 is 1/100, .01, or 1%. Fill in Data Table 1 for well three.
4) Keep following this same pattern and fill the other wells up to well 9. Make sure you do not mix the waters and pipettes, keep them in the same beakers.

Data Table 1

Well	Color Intensity	Concentration of parts of dye per part of the solution
1		
2		
3		
4		
5		
6		
7		
8		
9		

Questions:

1) Which is less concentrated, the solution in well one or well 2? How do you know?

2) Which well has a concentration of one part per million?

3) Which well number did the solution first appear colorless.

   a. What is the concentration in this cup?

4) Arsenic is poisonous in a concentration above 0.2 ppm (parts per million). Which is the first well to leave a concentration less than a concentration considered poisonous?

5) Fluoride is good for our teeth, but too much concentration can poison us and make us sick. How has our dental health improved by putting fluoride into the water, but at levels that won't make us sick?

6) Chlorine is poisonous to humans but is put in water to kill any possible microbes that could make us sick. How can we still have healthy water and chlorine in it?

7) What could be some sources of error in this investigation?

8) What careers do you think would be concerned with water pollution?

The Effect of Acid Deposition on Aquatic Ecosystems

Directions:

You will need two **small beakers**, two **water samples**, **vinegar**, a **conductivity probe**, and a **pH probe** attached to an **interface** connected to a **computer** with **Logger Pro**. Looking at the materials and lab we will be using, what are the safety precautions we should take to protect ourselves and materials during the investigation?

1) Take your two water samples, put 50 mL of each in separate beakers, and label them.
2) Use your conductivity sensor to find the conductivity of each water sample. Make sure the units are µS/cm. Use Data Table 1 with this measurement to determine each water sample's hardness and put this information in Data Table 2.
3) Find the initial pH of your first water sample and place that measurement in Data Table 2. You may have to stir your pH meter for a while until the reading levels out. (Tap water will always be just over 7.)
4) Add one drop of vinegar to the water sample and find the new pH measurement. Add this measurement to Data Table 2.
5) Repeat the procedure in #4 until you have added five drops of vinegar to the first sample.
6) Make sure you rinse off your pH sensor before testing the second sample.
7) Then repeat the same procedure you just did with the first water sample in #s 3-5 for the second water sample. Make sure you stay patient with your pH meter and wait for it to level off. Then rinse off the pH sensor when done.

Data Table 1

Conductivity (µS/cm)	Hardness
0-140	Very Soft
140-300	Soft
300-500	Slightly Hard
500-640	Moderately Hard
640-840	Hard
Above 840	Very Hard

Data Table 2

Sample 1, Source:	Sample 2, Source:
Conductivity:	Conductivity:
Hardness:	Hardness:
Initial pH:	Initial pH:
pH after 1 drop:	pH after 1 drop:
pH after 2 drops:	pH after 2 drops:
pH after 3 drops:	pH after 3 drops:
pH after 4 drops:	pH after 4 drops:
pH after 5 drops:	pH after 5 drops:

Questions:

1) How much did the pH drop for Sample 1?

2) How much did the pH drop for Sample 2?

3) How does the harness seem to affect the pH change of the water?

4) Why do you think we keep tap water harder than other freshwater?

5) How do you think fresh water in nature will be affected by acid rain compared to tap water?

Soil and Acid Rain

Directions:

You will need a **soil sample**, a **sand sample**, cut off **tops of two 2 liter plastic bottles**, two **coffee filters**, four **250 mL beakers**, **500 mL of a mixture of water and vinegar** in a **beaker**, and a **pH probe** attached to an **interface** connected to a **computer** with **Logger Pro**. **Looking at the materials and lab we will be using, what are the safety precautions we could take to protect ourselves and materials during the investigation?**

1) Take the cut-off top of a two-liter bottle and place a coffee filter in the bottle to cover the bottle's opening, forming a bowl in that bottle. Then fill the soil sample up to the top of the coffee filter in the bottle, but do not go over the filter's top. Place this setup into a 250 mL beaker where the bottle's opening is inside the beaker.
2) Rinse your pH probe and place it into the mixture of water and vinegar. Gently stir the probe in the mixture until the pH levels out (this takes lots of patience). Write this data in Data Table 1.
3) Slowly pour about 100 mL of the water and vinegar mixture into the soil and let it drain into the beaker. Ensure the mixture goes down into the soil and does not run off to the edge of the filter. If none of the mixtures comes out into the beaker, add more until it does.
4) Repeat the procedures in #s 1-3 for a sample of sand.
5) Rinse your pH probe and place it into the drained water and vinegar mixture to find the new pH. Write this data into Data Table 1 for both the soil and sand samples.

Data Table 1

	Soil Sample	Sand Sample
The pH of the mixture before it filtered through the soil		
The pH of the mixture after it filtered through the soil		
Change in pH		

Questions:

1) Did the mixture get more acidic or less acidic for each sample?

2) Which sample caused the biggest change in pH?

3) A buffer is a substance that helps neutralize a mixture. Which of the two soil samples is a better pH buffer?

4) If you lived in an area with acid rain, which of the two soil samples would you rather have in your yard and why?

5) What could be some sources of error in this investigation?

Investigating Salinity

Directions and Questions:

You will need to collect a water sample from a body of **saltwater**, like an estuary or an ocean. You will also need a **beaker** to put the water in and a **conductivity probe** attached to an **interface** connected to a **computer** with **Logger Pro**. **Looking at the materials and lab we will be using, what are the safety precautions we should take to protect ourselves and materials during the investigation?**

1) Make sure your conductivity probe switch is on the 0-20000 µS/cm salinity range, and the units are set to ppt. Put your probe in the water and find your water sample's salinity to the ppt (parts per thousand). What is the salinity of your sample?

2) How does your salinity sample compare to an average ocean salinity of 35 ppt?

3) How could the salinity change in the ocean?

4) How could the salinity change in estuaries?

Basic Information about Estuaries

Directions:
Go to: https://www.epa.gov/nep/basic-information-about-estuaries to help you answer the questions below.

1) What is an estuary?

2) Why are estuaries important?

3) How are estuaries critical to habitats?

4) How do estuaries have economic value?

5) What environmental services do estuaries perform?

6) How do estuaries act as protective buffers?

7) How are estuaries threatened?

Water Treatment

Directions:

Use the **internet** and your **textbook** to research the information needed to answer the questions below.

1) Where is the source of water that supplies your home?

2) Is it surface or groundwater?

3) How is it made safe for you to use?

4) Where does the water go when it leaves your house?

   a. Is it recycled and used again? If so, how is it used again, and where is it used again?

5) How is it made safe to go back into the environment?

Water Treatment Testing

Directions:

You will need a **beaker of tap water**, a **pH probe**, a **conductivity probe,** and a **turbidity probe** attached to an **interface** connected to a **computer** with **Logger Pro**. You will also need **distilled water** to help calibrate the turbidity probe and the internet for research. **Looking at the materials and lab we will be using, what are the safety precautions we should take to protect ourselves and materials during the investigation?**

1) Rinse off your pH probe and then place it in your water sample and gently stir. The pH probe will take some time to even out, so have some patience. Once it levels out, compare your measurement to the EPA Standards seen in Data Table 1. Then fill out the right side of Data Table 1 for pH.
2) Take your conductivity probe and place it into the water sample. Make sure the units for the probe are set for mg/L. Compare your measurement to EPA Standards seen in Data Table 1. Then fill out the right side of Data Table 1 for Total Dissolved Solids using the measurement you got from the conductivity probe.
3) Turbidity sensors need to be calibrated each time they are used. Use the instructions that come with the probe to calibrate it before using the probe. Then place your water sample in the vial at the appropriate level (at the white line). Ensure the outside is wiped clean with lens paper and there are no bubbles in the water sample. If either of these happens, you will have an error in your reading. Compare your measurement to the EPA Standards in Data Table 1. Then fill out the right side of Data Table 1 for turbidity.

Data Table 1

Contaminant	EPA Standard	Your Measurement	Did it meet EPA Standards?
pH	6.5-8.5		
Total Dissolved Solids	< 500 mg/L		
Turbidity	< 5 NTU		

Questions:

1) Which of the measurements met the EPA Standards?

2) Which of the measurements did not meet the EPA Standards?

3) Use the internet or your textbook to research how the unsettled particles that give the water an unpleasant, murky appearance are removed from water?

4) Why do most communities adjust the pH of tap water?

5) What could be some sources of error in this investigation?

6) What careers would do water treatment testing?

Quality of a Body of Water

Directions:

You will need **lens paper**, a **beaker** of a sample of a **local body of water**, a **temperature probe**, a **pH probe**, a **conductivity probe**, and a **turbidity probe** attached to an **interface** connected to a **computer** with **Logger Pro**. You will need **distilled water** to help calibrate the turbidity probe. **Looking at the materials and lab we will be using, what are the safety precautions we should take to protect ourselves and materials during the investigation?**

1) Place your temperature probe in your water sample and wait for it to level out. Compare your measurement to the standards in Data Table 1. Then fill out the right side of Data Table 1 for Temperature.
2) Rinse off your pH probe and then place it in your water sample and gently stir. The pH probe will take some time to even out, so have some patience. Once it levels out, compare your measurement to the standards seen in Data Table 1. Then fill out the right side of Data Table 1 for pH.
3) Take your conductivity probe and place it into the water sample. Make sure the units for the probe are set for mg/L. Compare your measurement to the standards seen in Data Table 1. Then fill out the right side of Data Table 1 for Total Dissolved Solids using the measurement you got from the conductivity probe.
4) Turbidity sensors need to be calibrated each time they are used. Use the instructions that come with the probe to calibrate it before using the probe. Then place your water sample in the vial at the appropriate level (at the white line). Ensure the outside is wiped clean with lens paper and there are no bubbles in the water sample. If either of these happens, you will have an error in your reading. Compare your measurement to the standards in Data Table 1. Then fill out the right side of Data Table 1 for turbidity.

Data Table 1

Contaminant	Standard for most freshwater life	Your Measurement
Temperature	5-25°C	
pH	6.5-8.2	
Total Dissolved Solids	< 1000 mg/L	
Turbidity	< 100 NTU	

Questions:

1) How does your water sample compare to the standards?

2) Would your water sample be suitable for most life?

3) What are some natural causes of the change in temperature in bodies of water?

4) What thermal pollution sources change the water temperature, thus changing indigenous organisms' habitat in that water?

5) What are some things that can change the pH of water in nature?

6) What are some causes for water to become more or less turbid?

What is your Watershed?

Directions:

Use the **internet** to find your county's website for your local watersheds. Research the information on the site to answer the questions below for your watershed.

1) What is a watershed?

2) How many watersheds are in your county?

3) What is your watershed?

4) Are there any creeks or reservoirs located within your system?

5) If so, how many?

6) What is the watershed for your school?

7) Find your home. Do you live in a flood plain? If so, how many years is it?

8) Flood insurance is something everyone can buy, but it costs more if you live in a flood plain. It also covers your house if a water heater or a pipe bursts in your home. Most home insurance policies do not cover this. Do you think you want to get flood insurance if and when you buy a home?

9) How close is the nearest flood plain to your home, and what kind is it?

Convection in Liquids

Directions and Observations:

1) Fill a **large beaker** with **water**, add **pepper** to it, place it on a **hotplate,** and heat the water to just below the boiling point with your **safety goggles** on. **Looking at the materials and lab we will be using, what are the safety precautions we should take to protect ourselves and materials during the investigation?**

2) Draw a picture of the motion of the pepper in hot water:

3) Describe the motion you see in hot water:

4) What do you think is causing this motion (go into detail)?

Questions:

1) Describe how the particle of the pepper moved as the water became hotter.

2) Describe how the particles of pepper moved as the water became colder after losing heat on the surface.

3) Explain how convection currents formed in the beaker.

4) Explain why the motion of the particles changed as the burner heated up the water.

5) How can this information show how currents can form in large bodies of water, including oceans?

Human Dependence and Influence on Oceans

Directions:

Use the **internet** to help you answer the questions below. Use your teacher's instructions to look for reputable sites that give you unbiased, correct information.

1) What are the causes of ocean currents?

   a. How is thermohaline circulation involved?

2) How does the Gulf Stream bring heat up to Europe?

3) What happens when that heat pump is turned off?

4) How was it turned off in the past?

5) Where do most of the oxygen come from for photosynthesis?

6) Where do most of the world's fish come from to feed people?

7) How do oceans affect weather for people onshore?

8) How do the oceans cause coastal winds and clouds?

9) Where do hurricanes happen the most? Explain why.

10) Where are hurricanes occurring more? Explain why.

11) How can humans produce electricity from oceans?

12) Where does all the runoff eventually flow to?

13) What happens if humans put plastics on the ground and runoff gets ahold of them?

   a. Where do we see the results of this?

14) When the polar ice melts, where does that melt go?

15) How high can it raise the ocean level?

16) How does the melt of the ice caps affect coastal cities?

    a. Which cities have been flooded by the rise of the ocean level?

    b. Which islands have been covered by the rise of the ocean levels?

17) How are humans causing sea levels to rise?

18) Wherever humans pump oil from the ocean bottom, what will happen every time?

    a. Where could that oil end up because of the ocean currents?

19) Draw a picture of the ocean currents that flow around the world. Is any part of the world unaffected by another part of the world?

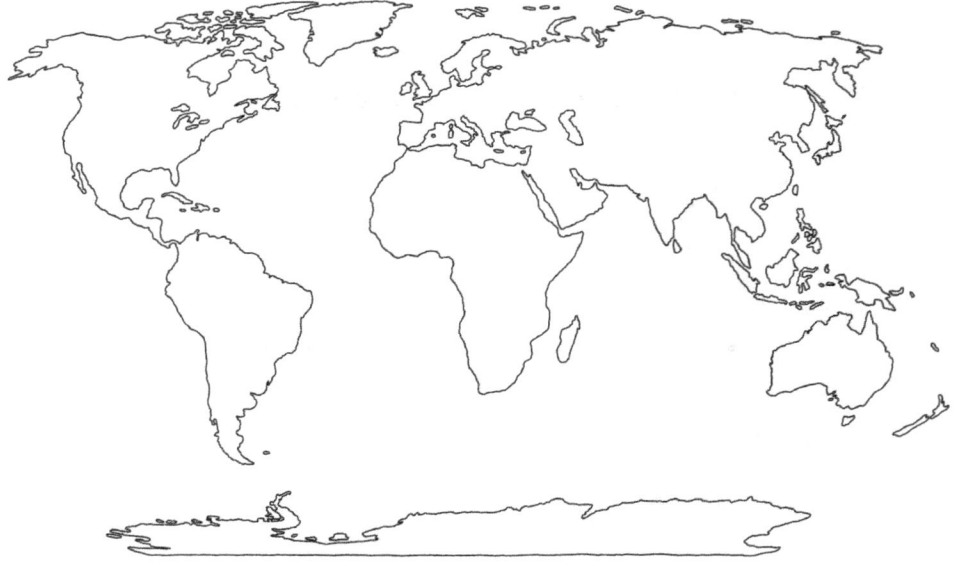

20) When radiation leaked into the ecosystem from a nuclear reactor meltdown in Japan, how did that affect Americans?

21) What are some causes of transgressive sedimentary sequences?

    a. How can this affect humans?

    b. Give examples of some that have happened in the past.

22) What are some causes of regressive sedimentary sequences?

    a. How can this affect humans?

    b. Give examples of some that have happened in the past.

The Great Pacific Garbage Patch

Directions:

Use the **internet** to research the Great Pacific Garbage Patch and answer the following questions.

1) Where is the Great Pacific Garbage Patch?

2) How big is it?

3) What is the source of the garbage?

4) What materials are in the garbage?

5) What is causing garbage to collect in this area?

6) What effect is this garbage having on the ecosystem there?

7) What is happening under the surface of the ocean there?

8) What legislation and efforts have been put in place to combat this problem?

9) What are efforts helping clean it up?

10) What are the difficulties in trying to clean up the garbage patch?

11) What is something we can do to help combat this situation?

Virtual Investigations that go with Human Effects on Hydrosphere

ExploreLearning.com:

   Water Cycle

   Water Pollution

   Coral Reef 1 – Abiotic Factors

   Coral Reef 2 – Biotic Factors

Unit 10: Influences to Climate Change

Sun → solar radiation

- solar radiation → atmospheric patterns
- solar radiation → regional changes → **Climate Change**
- climate factors
- ocean circulation patterns → atmospheric changes

atmospheric patterns → Weather Patterns (hurricanes, clouds, seasons, wind)

Human Activities → atmospheric patterns
Human Activities → Regional and Global Consequences

Natural Events / Human Activities:
- volcano
- fires
- meteor impacts
- pollution emissions
- burn fossil fuels
- population growth
- municipal development

→ air quality → atmospheric pollution → Climate Change

Regional and Global Consequences → **Ecological Consequences**
- acid rain
- ozone depletion
- smog
- increase greenhouse gases

285

Composition of the Atmosphere

1) Use the **internet** to list the top 5 chemicals and their % in the atmosphere (include how much the water range can be).

2) Describe the Earth's first atmosphere and tell how it may have changed to what it is today. **Hint:** There were 3 atmospheres. Tell what the first atmosphere was composed of, then what happened to cause the second, then what happened to cause the one we have today.

3) When nature spent billions of years locking the old atmospheric chemicals in the Earth's crust, what do you think would be the effects of digging them up and burning them, putting them back into the atmosphere?

The Greenhouse Effect on Earth

Directions:

You will need two **plastic tubs** (shoebox size) **painted black** on the inside, a **"Press'n Seal"** **sealing wrap**, two **temperature probes** attached to an **interface** connected to a **computer** with **Logger Pro**, and a **light source** like an incandescent lamp or some other lamp that gives off heat. **Looking at the materials and lab we will be using, what are the safety precautions we should take to protect ourselves and materials during the investigation?**

1) Make sure your tubs are both painted black on the inside, and a hole is poked through the end of each tub, big enough for a temperature probe to fit in snugly. Push the temperature probes through the holes of both tubs.
2) On the first tub, make sure the Press'n Seal is secured to the tub's opening and the temperature probe is plugged into channel 1 of the interface.
3) The other tub needs to remain open, and the temperature probe plugged into channel 2.
4) In Logger Pro, open the folder Earth Science with Vernier and file #24 Greenhouse Effect.
5) **Hypothesis:** which tub do you think will heat up faster?

6) Make sure your light source is equal distance from both tubs. Press "Collect" in the Logger Pro and turn on your lamp.
7) Monitor the time; when **5 minutes have passed**, **turn off the light**. Data will continue to be collected.
8) At **10 minutes, turn the lamp back on**. Data collection will continue until 15 minutes. At 15 minutes, Data collection will stop.
9) Look at the data collected in the Logger Pro and fill in Data Table 1 for the initial temperature, the temperature at 5 minutes, the temperature at 10 minutes, and the temperature at 15 minutes.
10) Then subtract the temperatures between Probe 1 and Probe 2 to get the temperature differences at different times. Write this information on the right side of Data Table 1.

Data Table 1

	Probe 1 Greenhouse	Probe 2 Control	Temperature Difference
0 Minute Temperature (°C)			
5 Minute Temperature (°C)			
10 Minute Temperature (°C)			
15 Minute Temperature (°C)			

Questions:

1) When the Lamp was on, which tub heated faster?

2) Give a possible explanation for your answer to number 1.

3) When the lamp was off, which tub cooled faster?

4) Give a possible explanation for your answer for number 3.

5) How does this information show how temperatures could increase over time because of the greenhouse effect?

6) What are we doing to the atmosphere to increase greenhouse gases' effects to cause global warming and climate change?

7) How could this information show temperature decrease over time without the greenhouse effect?

8) What do you think happened to the atmosphere to allow ice ages to happen on the Earth?

9) When do you think the temperature will get the coldest at night: when it is a cloudy night or when there are no clouds at night? Explain why.

10) The habitability of the Earth is a result of a delicate balance of the greenhouse effect. How/why is this statement true?

11) Explain why a closed automobile heats up in the sun.

12) Why do you not leave your child or pet in the car on warm days when the car is parked and turned off?

13) What could be sources of error in this investigation?

14) Do you think the Greenhouse Effect is a hypothesis or a theory? Explain why.

Climate and Greenhouse Gases: Data Table

Directions:

Look at the Data Table below and build line graphs on the following pages showing the trends in temperature and carbon dioxide data. Then compare the trends in the graphs together, answering the questions that follow. This data was collected through ice core samples where the atmosphere was trapped inside the snow and then compressed into ice. When we melt the different layers of ice, we can accurately measure Carbon Dioxide levels in those layers. Also, by measuring the number of Oxygen isotopes, we can get an accurate temperature of that time.

Data Table 1

Years Before Present (x 1000)	Local Temperature Change (°C)	Carbon Dioxide (ppm)
160	-9	190
150	-10	205
140	-10	240
130	-3	280
120	1	278
110	-4	240
100	-8	225
90	-5	230
80	-6	220
70	-8	250
60	-9	190
50	-7	220
40	-8	180
30	-7	225
20	-9	200
10	-2	260
0 (Year about 1850)	-0.5	280
0 (Year about 2002)	-	371

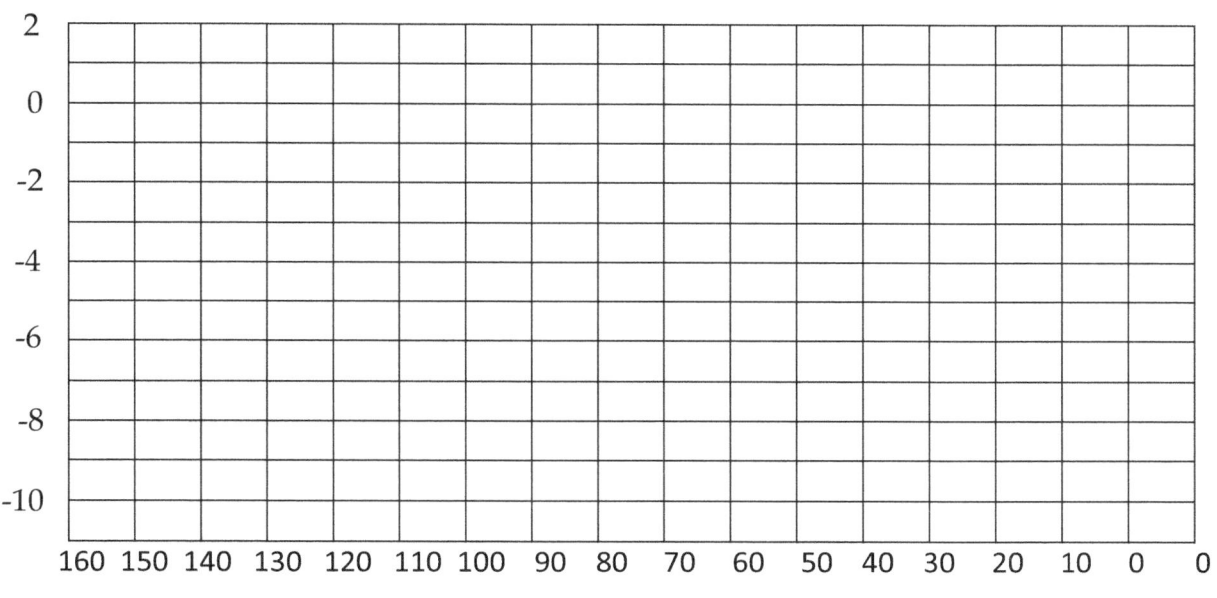

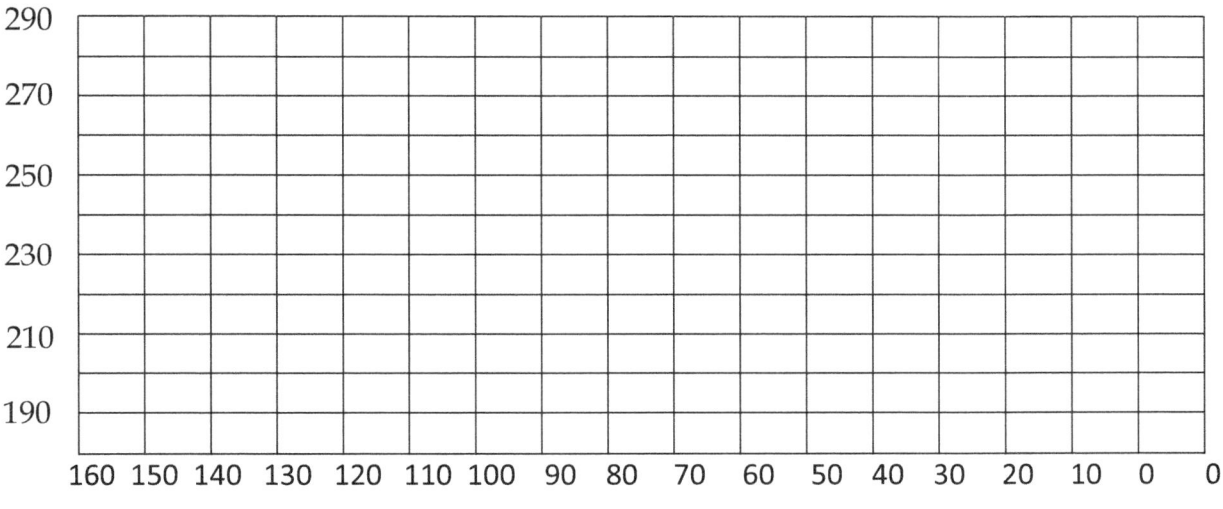

Questions:

1) Do you see any similarities between the temperature graph and the Carbon Dioxide levels graph?

2) Which graph literally went off the chart and where?

3) What could this mean for the other graph in the near future?

4) Does this evidence show Carbon Dioxide is causing the temperature change? Explain.

5) How might this affect our future climate and where people live if the Earth is getting warmer?

6) How could this affect the evolution of life?

7) Can anything be done to prevent increasing temperatures? If so, what are they?

8) Do you think Global warming is a hypothesis or a theory? Explain why?

Carbon Dioxide and Population

Directions:

Graph the information from Data Table 1 onto Graph 1, then answer the questions below. Use the left Y-axis to create a line graph for the Population and the right Y-axis to create a line graph for the carbon dioxide emissions. Make sure you make a key showing which line on the graph is the population and which line is the carbon dioxide emissions.

Data Table 1

Year	Population (in millions)	Carbon Dioxide Emissions (in metric tons)
1750	790	11
1800	980	29
1850	1260	198
1900	1650	1,982
1950	2520	5,982
2000	6060	25,620

Graph 1

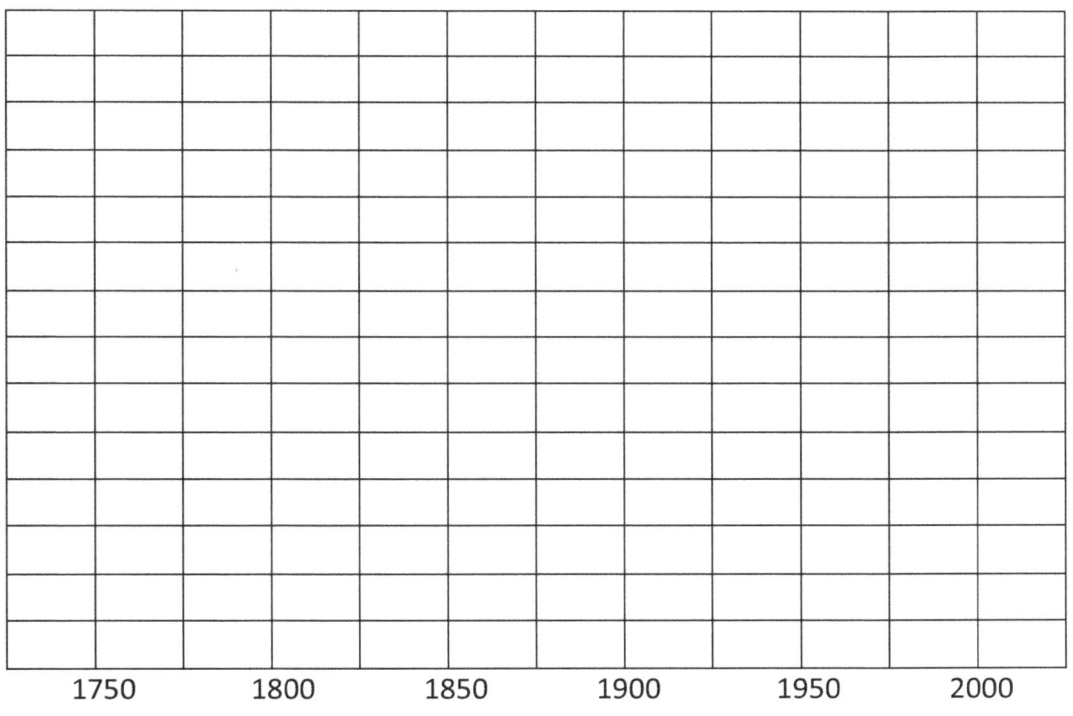

Questions:

1) What happens to the carbon dioxide emissions as the population of humans rises?

2) Why do you think this happens?

3) How has the source of carbon dioxide emissions changed over the years?

4) Is there any way we can lower those emissions? If so, how?

5) What kind of culture change do we need if we are going to lower the carbon dioxide levels?

6) What is happening to the Earth because of the rising carbon dioxide levels?

7) How does this impact humans now?

8) How could this impact humans in the future if we do not change the trends in the data?

9) How could this impact the evolution of life in the future if the trends continue?

Carbon Emissions

Directions:

1) Your teacher will assign you a family scenario. Use it as you follow the directions to fill in your Carbon Emissions in Data Table 1. Once done, check with another person with the same family as you to check for errors. Once done, follow the directions again to fill out a carbon emissions Data Table 1 for your family. Use a five days a week work week, 52 weeks a year, four weeks in a month, 12 months a year, and 365 days a year. Everyone will not use all parts of the calculation. Only use the ones that apply to your family.

2) To comply with the **Kyoto Protocol**, we should only produce 5.4 tons of CO_2 per person per year. To make a difference to help control global warming, we could produce no more than 2.35 tons of CO_2 per person per year.

3) Number of miles traveled by car #1 each year_____, divided by average miles per gallon = _____ gallons of gas multiplied by 22 pounds of CO_2/gallon of gas = _____ pounds of CO_2 from Car 1. Write this number for Car #1 in Data Table 1.

    a. Repeat #3 for additional cars and any other fuel motor vehicles, including motorcycles, boats, etc.

4) Number of miles of air travel per year (all household members) _____, multiplied by 0.9 pounds of CO_2/mile of air travel = _____ pounds of CO_2. Write this in Data Table 1.

5) Number of miles traveled on mass transit (bus, train) _____, multiplied by 0.5 pounds of CO_2/mile of mass transit travel = _____ pounds CO_2. Write this in Data Table 1.

6) The number of miles traveled by taxi, limo, UBER _____, multiplied by 1.5 pounds of CO_2/mile in a taxi, limo, or UBER = _____ ponds of CO_2. Write this in Data Table 1.

7) Kilowatt-hours of electricity used per year _____, multiplied by 1.5 pounds of CO_2/kilowatt-hour = _____ pounds of CO_2. Write this in Data Table 1.

8) Therms of natural gas per year _____, multiplied by 11 pounds CO_2/therm = _____ pounds of CO_2. Write this in Data Table 1.

9) Gallons of propane or bottled gas per year _____, multiplied by 13 pounds CO_2/gallon = _____ pounds of CO_2. Write this in Data Table 1.

10) Add up the pounds of CO_2 emitted by the household. Write this in Data Table 1.

11) Divide the total by the number of people in the household = _____ pounds of CO_2 emitted per person in one year. Divide the total by 2000 pounds to determine the number of tons of CO_2 per year per person. Write this in Data Table 1.

12) Find the tons of CO_2 per year per person for the other three families from the other groups in the class.

13) Repeat the procedure for #2-11 for your own family. Write this into Data Table 3.

Family #1

Ralph and Brenda are both in their early 30s and live inside the Loop in a 1200 square foot condo. They both work downtown and take the bus to work. She commutes about 8 miles per workday on the bus, and he travels about 7 miles per day on the bus. Since they do not need cars for a commute, they own one car, an Acura TL, that gets 23 miles per gallon. They drive the car on average 7 miles per day because they live close to dining and recreation. They fly an average of 3000 miles a year each to visit family and vacation for the year. Their condo uses an average of 950 kilowatts of electricity each month, and all appliances, including heating and air conditioning, are electric.

Family #2

Bill and Ellen are in their early 40s with one child, Mark. They live in the Heights in an 1800 square foot home that they remodeled to be energy efficient. They are close to everything they need, including a grocery store and a farmer's market. Both work; however, they have an office inside the house, allowing Bill to work from home three days a week. The other two days a week, he commutes to his nearby office. Ellen is an artist who works out of a loft in their house. They own energy-efficient cars. He drives a Honda Civic Hybrid that gets about 40 miles to the gallon, and she drives a Toyota Prius that gets 45 miles to the gallon. On average, Bill drives 30 miles each week, and Ellen drives 40 miles each week. Mark attends a nearby school and rides his bike to school. Despite their house being fully electric, they only use an average of 350 kilowatts of electricity per month because they have solar panels on the roof of their house that generate part of the electricity for their household.

Family 3

Matt and Debbie, who are in their 40s, live in the suburbs with their three children. Their house is 2000 square feet and uses all electricity. On Average, they use 1200 kilowatts of electricity per month. Debbie works part-time and drives a minivan that gets 20 miles per gallon. Between work, running family errands, and driving children, she drives about 30 miles per day. Matt has to drive more for his work and sometimes drives to other work sites, so he puts about 60 miles per day on his pick-up truck that gets 14 miles to the gallon. To relax, Matt enjoys grilling on their outdoor propane grill. On average, he uses 1.5 gallons of propane each month.

Family 4

Ben and Dawn live in a 4,000 square-foot house with all the amenities, including a heated pool, a family room with a built-in home theater system, and several televisions throughout the house, including one upstairs solely for the kids' Wii. With a house this large and with so many electrical devices, they use on average 5000 kilowatts of electricity per month. The house has a natural gas stove and oven, outdoor grill, water heater, and furnace. These appliances use on average 65 therms of natural gas each month. Dawn drives a full-sized SUV that gets 18 miles to the gallon, which she uses to take the kids to their private school, run errands for herself and her family, and take kids to their after-school activities. On average, Dawn drives about 60 miles per day. Ben enjoys driving his Hummer H3, which gets about 13 miles to the gallon. He has a fairly long commute to work, so he drives about 50 miles per day. Twice a year, they go on vacation as a family. Each family member travels by airplane an average of 7000 miles each year for these vacations. Ben has to travel for business once a month. He takes a taxi to and from the airport, which is 25 miles from their house. He flies to Boston, which is 1600 miles away.

Data Table 1: Family #_____

Source per Year	Car 1	Car 2	Air Travel	Mass Transit	Taxi UBER	KWh	Therms Natural Gas	Propane	Total	Total per Person	Tons per person
Pounds CO_2											

Data Table 2: Family Comparison

Family	Family 1	Family 2	Family 3	Family 4
Tons of CO_2 per Person				

Data Table 3: My Family

Source per Year	Car 1	Car 2	Air Travel	Mass Transit	Taxi UBER	KWh	Therms Natural Gas	Propane	Total	Total per Person	Tons per person
Pounds CO_2											

Questions:

1) What was the largest contributor to the family you were assigned carbon emissions?

2) Would this family's habits help or hinder efforts to control global warming?

3) Compare this family to the other families. What are the biggest differences?

4) Compare your family in Data Table 3 to families 1-4. Which family does your family closely resemble?

5) Is your family Kyoto compliant?

6) Would your family help or hinder efforts to control global warming? Explain.

7) What are some things your family could do to help efforts to control global warming?

Climate Change

Directions:

Use the **internet** to research climate change and answer the questions below.

1) What is climate change?

2) What are the causes?

3) What role does plate tectonics play in climate change?

4) What is the impact of climate change on polar ice caps and glaciers?

5) How would this affect ocean currents?

6) If ocean temperatures keep rising, what are the consequences for life in the ocean?

7) How do ocean currents affect the surface temperatures of the Earth?

8) How is it impacting humans worldwide, and what can we expect in the future if this is happening?

9) What evidence do we have that it is happening?

10) What are the arguments against climate change?

11) Who are the people (occupations) studying climate change?

12) Do you think it is man-made or natural? Explain in detail why with evidence for your opinion.

Climate Change Poster

Directions:

Build a poster promoting awareness of Climate Change and Global Warming, showing some causes, effects, and possible solutions people could do to make it better. Do a rough draft here and share your ideas with your teacher or classmates to get feedback. Then produce either a hardcopy or digital poster according to your teacher's instructions.

Environmental Issues and Ethics

Directions: Use the **internet** and **textbook** to help you fill in the chart about environmental policies.

Legislation, Treaty, or Protocol	What is it about?	Who enacted it?	Where is it in effect?	Why was it made?
Clean Air Act				
Clean Water Act				
Soil and Water Resources Conservation Act				
Texas Automobile Emissions Regulations				
National Park Service Act				

Endangered Species Act				
Environmental Arctic Treaty System				
Montreal Protocol				
Kyoto Protocol				
Paris Accord				

1) What do you think is the purpose of laws, treaties, and protocols?

2) What careers would be interested in these laws, treaties, and protocols?

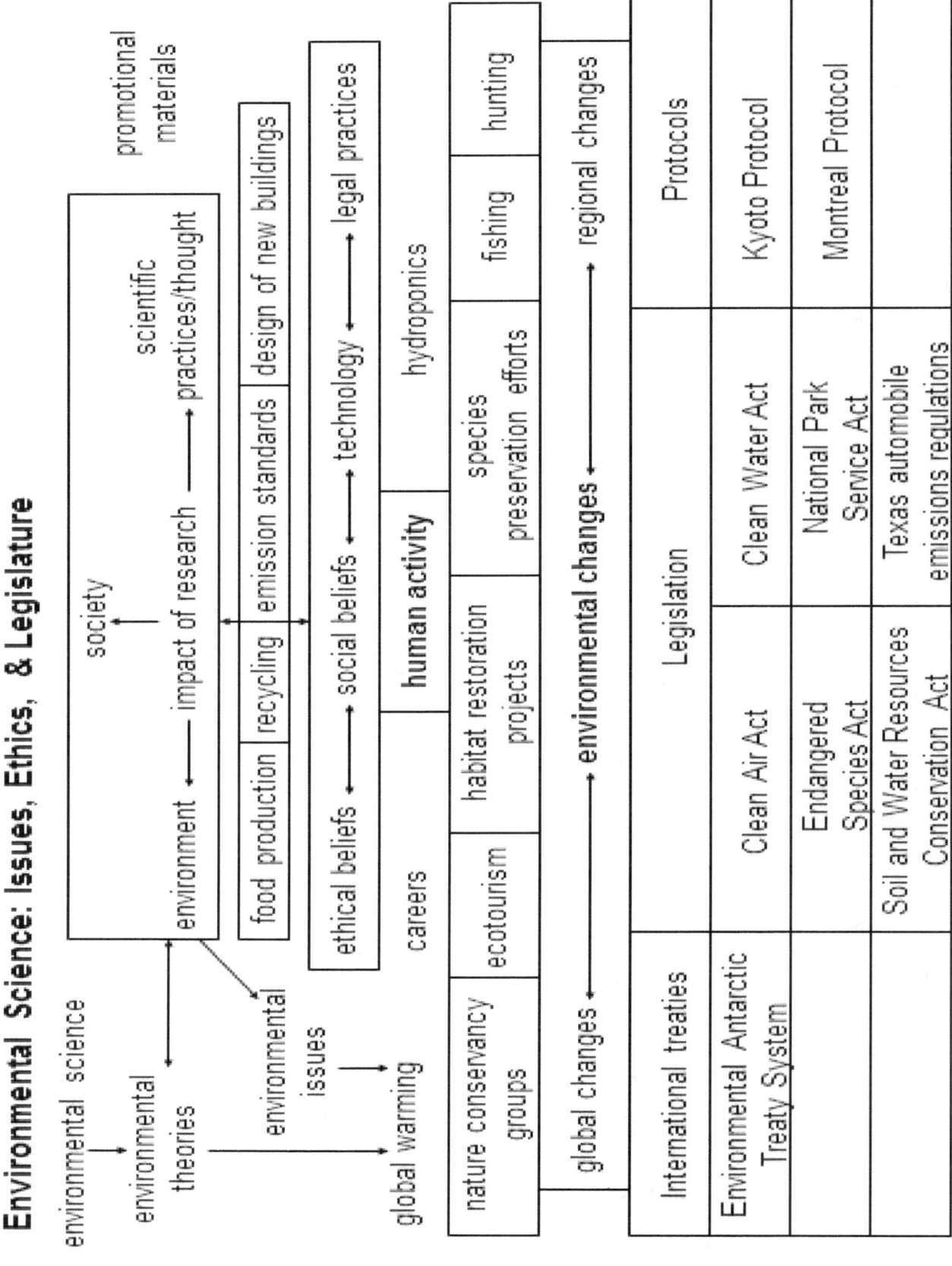

Virtual Investigations that go with Influences of Climate Change

ExploreLearning.com

Greenhouse Effect

Coral Reefs 1 – Abiotic Factors

Coral Reefs 2 – Biotic Factors

Ocean Carbon Equilibrium STEM Case

Ocean Carbon Equilibrium Handbook

Seasons Around the World

Observing Weather

Weather Maps

Convection Cells

Weathering

Earth and Space Science TEKS/NGSS Correlations

Focus on the Process Earth c 1ABCDEG 2B 3AB 4A

Measurement Lab Earth c 1ABCDEF 2C 3AB

Patterns in Pennies Earth c 1ABCDEF 2BC 3AB 4A

Virtual Investigations that go with Introduction Earth c 1ABEFGH 2ABCD 3AB 4ABC

The Development of the Universe Concept Map Earth c; HS-ESS1-123

Star's Life Cycle Earth c 1ABE 3B 4A; HS-ESS1-3

Star Life Cycle Model Earth c 1ABCDEG 2ABD 3AB; HS-ESS1-3

Our Bright and Mourning Star Earth c 1ABE 3AB 4AB; HS-ESS1-123

Nuclear Fission and Fusion in a Star Earth c 1ABEFG 3AB 4AB; HS-ESS1-1

Compare and Classify Stars Earth c 1ABEG 2B 3AB 4A; HS-ESS1-23

Classifying Galaxies Earth c 1ABEG 3AB 4AB; HS-ESS1-2

A Guide to the Milky Way Galaxy Earth c 1ABEFG 2A 3AB 4AB; HS-ESS1-2, MS-ESS1-2

Ten Things NASA Wants you to Know about the Universe Earth c 1ABEG 2B 3AB 4AB; HS-ESS1-2

The History of the Big Bang Theory Earth c 1ABEFGH 3AB 4AB; HS-ESS1-2

The Pixel of the Universe Earth c 1ABCDEGH 2AB 3ABC 4A; HS-ESS1-2

WMAP Earth c 1ABEG 2B 3AB 4A; HS-ESS1-2

Virtual Investigations that go with the Development of the Universe Earth c 1ABEFGH 2ABCD 3ABC 4ABC; HS-ESS1-123

Characteristics of the Solar System Concept Map Earth c 5B; HS-ESS1-4, MS-ESS1-12

Simulating the Orbit of a Planet and Sun Earth c 1ABCDEG 3AB 4A; HS-ESS1-4, MS-ESS1-2

Bending of Space-time Earth c 1ABCDEG 2ABD 3AB 4A; HS-ESS1-4, MS-ESS1-2

Kepler's Laws Earth c 1ABE 3AB; HS-ESS1-4, MS-ESS1-2

The Push of Gravity Earth c 1ABE 2BC 3AB 4A; HS-ESS1-4, MS-ESS1-2

Observing Forces in Orbit (a) Earth c 1ABCDEG 2A 3AB 4A; HS-ESS1-4, MS-ESS1-2

Solar System NASA Web Quest Earth c 1ABE 3AB 4A 5B; HS-ESS1-4, MS-ESS1-12

Scale Model of the Solar System Earth c 1ABCDEG 2ABC 3AB 4A; HS-ESS1-4, MS-ESS1-123

Scale Model of a Planetary System Earth c 1ABCDEG 2ABC 3AB 4A; HS-ESS1-4, MS-ESS1-123

Map of our Solar System Earth c 1ABEG 2AB 3AB; HS-ESS1-4, MS-ESS1-123

Evolution of our Solar System Earth c 1ABE 3AB 4A 5A; HS-ESS1-46, MS-ESS1-4

How we know the History of the Earth Earth c 1ABE 3AB 4A 7A; HS-ESS1-56, MS-ESS1-4

The Greenhouse Effect on Different Planets Earth c 1ABCDEFGH 2ABCD 3AB 4A 11CF; HS-ESS1-4,2-24, MS-ESS1-3

Seeing Patterns in Layers of the Atmosphere Earth c 1ABEF 2ABC 3AB 4A 11AB; HS-ESS2-2, MS-ESS1-3

Composition of Planet Atmospheres Earth c 1ABE 3B; HS-ESS1-4, 2-24, MS-ESS1-3

Atmospheres of Moons Earth c 1ABE 3AB 4A; HS-ESS1-4, 2-24, MS-ESS1-3

Models of Micro-molecules Earth c 1ABCDEG 2AB 3AB 4A; HS-ESS2-2, MS-ESS1-3

Building a Model of a Water Molecule Earth c 1ABCDEG 2AB 3AB 4A; HS-ESS2-2, MS-ESS1-3

Models of Macromolecules Earth c 1ABCDEG 2AB 3AB 4A; HS-ESS2-6, MS-ESS1-3

How is Life Allowed on Earth? Earth c 1ABE 3AB 4AB 6D; HS-ESS1-346, 2-1246, MS-ESS1-4

Virtual Investigations that go with the Characteristics of our Solar System Earth c 1ABEFG 2ABCD 3ABC 4ABC; HS-ESS1-4, 2-24, MS-ESS1-1234

Sun Earth and Moon Interactions Concept Map Earth c 5BC; HS-ESS1-4, 2-4, MS-ESS1-12

Seasons and the Tilt of the Earth Earth c 1ABCDEFG 2BC 3AB 4A 11ACF; HS-ESS1-4, 2-4, MS-ESS1-1

The Pull of Gravity on Earth Earth c 1ABE 2BC 3AB 4A 5B; HS-ESS1-4, MS-ESS1-12

Gravitational Interactions of the Sun and Moon on the Earth Earth c 1ABCDEG 2AB 3AB 4A 5B; HS-ESS1-4, MS-ESS1-12

Observing Forces in Orbits (b) Earth c 1ABCDEG 2ABD 3AB 4A 5B; HS-ESS1-4, MS-ESS1-12

Orbit Under Glass Earth c 1ABCDEG 2B 3AB 4A 5B; HS-ESS1-4, MS-ESS1-12

Moon Phases Earth c 1ABCDEFG 2ABD 3AB 4A 5B; HS-ESS1-4, MS-ESS1-1

Origins of the Moon Earth c 1ABE 3AB 4A 5C; HS-ESS1-6, MS-ESS1-4

Solar Eclipses Earth c 1ABCDEG 3AB 4A 5B; HS-ESS1-4, MS-ESS1-1

Solar Eclipse on Earth Perspective Earth c 1ABCDEG 3AB 4A 5B; HS-ESS1-4, MS-ESS1-1

Lunar Eclipses Earth c 1ABCDEG 3AB 4A 5B; HS-ESS1-4, MS-ESS1-1

Virtual Investigations that go with Sun Earth and Moon Interactions Earth c 1ABEFGH 2ABCD 3ABC 4ABC 5B; HS-ESS1-4, 2-4, MS-ESS1-12

Earth's Structure and Functions Concept Maps Earth c 8ABDEG; HS-ESS1-5, MS-ESS1-3,2-13

Observing the Different –spheres Earth c 1ABE 3AB 4A 6E; HS-ESS2-4567

Modeling the Layers of the Earth Earth c 1ABCDEG 3AB 4A 8BCD 6A 7A; HS-ESS1-5, 2-3, MS-ESS1-3

Soft Boiled Egg Earth Earth c 1ABCDEG 2AD 3AB 4A 8B; HS-ESS1-5, 2-3, MS-ESS1-3

Modeling the Rock Cycle Earth c 1ABCDEG 2AD 3AB 4A 8A; HS-ESS1-5, 2-3, MS-ESS2-1

Nuclear Decay Half-life of Pennies (Fossils) Earth c 1ABCDEF 2ABCD 3AB 4A 7A; HS-ESS1-6, MS-ESS1-4

Calculating Nuclear Half-life Decay (Fossils) Earth c 1ABEF 2BC 3AB 4A 7A; HS-ESS1-6, MS-ESS1-4

Types of Soil Earth c 1ABE 3AB 9BC; HS-ESS2-1245

Determining Soil Type Earth c 1ABCDE 2B 3AB; HS-ESS2-167, 3-12

Soil Salinity Earth c 1ABCDE 2B 3AB 4A 9BC; HS-ESS2-1567

Soil pH Earth c 1ABCDE 2BC 3AB 4A 9B; HS-ESS2-1567

Timeline of Tectonic Plate Theory Earth c 1ABE 4AB 4AC 8BC; HS-ESS2-13, MS-ESS2-3

Plate Tectonics Lab (Eatable) Earth c 1ABCDEG 3AB 4A 8ADEG; HS-ESS2-13, MS-ESS2-3

Plate Tectonics Lab (Non-Eatable) Earth c 1ABCDEG 3AB 4A 8ADEG; HS-ESS2-13, MS-ESS2-3

Effects of Plate Tectonics Earth c 1AB 3B 8ADEH; HS-ESS2-123456, MS-ESS3-2

Calculating the Rate of Tectonic Plate Motion Earth c 1ABEG 2BC 3AB 8F; HS-ESS2-13

Topographical Map on Earth Earth c 1ABCDEG 2BC 3AB 4A 9A; HS-ESS2-1

Interpreting Topographical and Geologic Maps of Your State Earth c 1ABCDEG 3AB 4A 9A; HS-ESS2-1

Virtual Investigations that go with Earth's Structure and Functions Earth c 1ABEFGH 2ABCD 3ABC 4AB 8ABFGH 9A; HS-ESS2-123, MS-ESS1-3, 2-13

Weather Patterns Concept Map Earth c 11ABCDF; HS-ESS2-245, MS-ESS2-2456

The Water Cycle Earth c 1ABEFG 3AB 4A 11AF; HS-ESS2-5, MS-ESS2-45

How does Rain Form? Earth c 1ABCDE 2B 3AB 4A 11F; HS-ESS2-5, MS-ESS2-45

How Hurricanes Form Earth c 1ABE 2B 3AB 4AB 11CDF 12BC; HS-ESS2-245, MS-ESS2-45

Global Wind Movement Earth c 1ABEG 3AB 4A 10C 11BCDF; HS-ESS1-4, 2-4, MS-ESS2-56

Model Showing the Rotation of the Earth Stirring up the Atmosphere Earth c 1ABCDEG 3AB 4A 10C; HS-ESS1-4, 2-4, MS-ESS2-56

A Local Weather Study Earth c 1ABCDEF 2BC 3AB 4A 11ABDF; HS-ESS2-245, MS-ESS2-56

Relative Humidity Earth c 1ABCDEF 2ABCD 3AB 4A 11ABF; HS-ESS2-245, MS-ESS2-56

Temperature Inversions Earth c 1ABE 3AB 4A 11FG; HS-ESS2-245

Weathering Earth c 1ABCDEG 2AD 3AB 9B; HS-ESS2-15, MS-ESS2-2

Erosion Earth c 1ABCDEG 2AD 3AB 9BC 12E; HS-ESS2-15, MS-ESS2-2

Biome Research Report Earth c 1ABEF 6E; HS-ESS2-6

Biomes Chart Earth c 1ABEF 6E; HS-ESS2-6

Virtual Investigations that go with Weather Patterns Earth c 1ABEFGH 2ABCD 3ABC 4ABC 9D 11ABCDFG; HS-ESS2-2456, MS-ESS2-2456

Energy Flow and Cycles Concept Map Earth c 6E; HS-ESS2-67, MS-ESS2-1

Build an Ecosystem Earth c 1ABCDEFG 2BC 3AB 4A 6E; HS-ESS2-6, MS-ESS2-1

Making a Food Web Earth c 1ABCDEG 6E; HS-ESS2-6, MS-ESS2-1

Ecological Pyramid Earth c 1ABCDEG 2BC 6E; HS-ESS2-6, MS-ESS2-1

Symbiotic Relationships Earth c 1AB 3AB 4A; HS-ESS2-67

Competitive Relationships Earth c 1ABG 3AB 4A 6E; HS-ESS2-67

Hierarchical Organizations of Ecosystems Earth c 1ABG 3AB 4A; HS-ESS2-67, MS-ESS2-1

Conservation of Life: Photosynthesis and Respiration Earth c 1ABEG 2BC 3AB 4A 6E; HS-ESS2-6, MS-ESS2-1

Biodiversity in Ecosystems Earth c 1ABEF 2BC 3AB 4A 6E 12D; HS-ESS2-67

Virtual Investigations that go with Energy Flow and Cycles Earth c 1ABEFGH 2ABCD 3ABC 4ABC 6E 12D; HS-ESS2-67, MS-ESS2-1

Ecological Changes Concept Maps Earth c 7ABCDEF 12 F; HS-ESS2-7

The Story of Life Earth c 1ABGH 2BC 3AB 4A 7CF 12F; HS-ESS2-7, MS-ESS1-4, 2-3

Fossil Evidenceof Relative Dating Earth c 1ABEG 2B 3AB 4A 7ABDEF; HS-ESS2-7, MS-ESS1-4

Variation Within a Population Earth c 1ABCDEF 2BC 3AB 4A 12F; HS-ESS2-7

Goldfish Evolution Earth c 1ABCDEF 2BC 3AB 4A 12F; HS-ESS2-7

Changing Environment for Beads Earth c 1ABCDEF 2BC 3AB 4A 12F; HS-ESS2-7

Causes of Invasive Species Earth c 1ABE 3AB 4A 12F; HS-ESS2-7, 3-46, MS-ESS3-3

Humans Changing Ecosystems Earth c 1ABE 3AB 4A 12F; HS-ESS2-7, 3-46, MS-ESS3-3

What Happens to the Food Web? Earth c 1ABG 3AB 4A 12F; HS-ESS2-67, 3-46, MS-ESS3-3

Our Little Mountain Earth c 1ABE 3AB 4A 12F; HS-ESS2-7, 3-46, MS-ESS3-3

Primary or Secondary Succession? Earth c 1ABE 12F; HS-ESS2-7, MS-ESS3-3

Virtual Investigations that go with Ecological Changes Earth c 1ABEFGH 2ABCD 3ABC 4ABC 12F; HS-ESS2-67

Managing Natural Resources Concept Map Earth c 12EF 13A; HS-ESS3-123, MS-ESS3-1234

Soil Moisture Earth c 1ABCDEF 2BC 3AB 4A; HS-ESS3-1, MS-ESS3-34

Infiltration Rate and Water Holding Capacity Earth c 1ABCDEF 2BC 3AB 4A; HS-ESS3-12, MS-ESS3-134

Managing Garden Soil Moisture Earth c 1ABCDEF 2BC 3AB 4A; HS-ESS3-12, MS-ESS3-34

Organic Gardening and Hydroponics Earth c 1ABE 3AB 4C 6E 12E; HS-ESS3-123, MS-ESS3-34

Sustainability of Diets for a Growing Population Earth c 1ABE 3AB 4C 6E 12F; HS-ESS3-123, MS-ESS2-1, 3-34

Nonrenewable Resources Chart Earth c 1ABEF 3AB 4A 12E 13A; HS-ESS3-12, MS-ESS3-1

Renewable Resources Chart Earth c 1ABEF 3AB 4A 12E 13A; HS-ESS3-12, MS-ESS3-1

Nuclear Reactor Earth c 1ABEFG 3AB 4A 12E; HS-ESS3-12, MS-ESS3-12

Natural and Manmade Disasters Earth c 1ABEF 12AB; HS-ESS3-1, MS-ESS3-23

Worldwide Disaster Presentations Earth c 1ABE 12AB; HS-ESS3-1, MS-ESS3-23

How to Dispose of Waste Earth c 1AB 3AB 4ABC 9D 12E 13AB; HS-ESS3-24, MS-ESS3-34

Waste Management Regulations Earth c 1AB 3AB 4ABC 9D 12E 13AB; HS-ESS3-24, MS-ESS3-34

Waste Management Plan Earth c 1AB 3AB 4AB 9D 12E 13A; HS-ESS3-24, MS-ESS3-34

Local Clean Up Earth c 1AB 3AB 4ABC 9D 12E 13AB; HS-ESS3-24, MS-ESS3-34

Virtual Investigations that go with Managing Natural Resources Earth c 1ABEFGH 2ABCD 3ABC 4ABC 12EF 13A; HS-ESS3-12, MS-ESS3-1234

Human Impact on the Hydrosphere Concept Map Earth c 9D 10ABC 11F; HS-ESS3-34, MS-ESS3-34

The Water Cycle Earth c 1ABEFG 3AB 4A 11AF; HS-ESS2-5, 3-34

Groundwater Pollution Lab Earth c 1ABCDEF 3AB 4AC 9BD 13B; HS-ESS3-3, MS-ESS3-34

Parts Per Million Earth c 1ABCDEF 3AB 4AC 9BC 13B; HS-ESS3-34, MS-ESS3-34

The Effect of Acid Deposition on Aquatic Ecosystems Earth c 1ABCDEF 3AB 4A 9BD; HS-ESS3-3, MS-ESS2-1

Soil and Acid Rain Earth c 1BCDEF 3AB 4A 9BD; HS-ESS3-3, MS-ESS2-1

Investigating Salinity Earth c 1ABCDE 3AB 4A 9BD; HS-ESS2-5, 3-3, MS-ESS2-6

Basic Information about Estuaries Earth c 1ABE 3AB 4AB 9BD; HS-ESS3-34, MS-ESS2-6

Water Treatment Earth c 1ABE 3AB 4A 9BD 13A; HS-ESS3-4, MS-ESS3-34

Water Treatment Testing Earth c 1ABCDE 2AD 3AB 4AC 9BD 13AB; HS-ESS3-4, MS-ESS3-34

Quality of a Body of Water Earth c 1ABCDEF 2B 3AB 4A 13A; HS-ESS3-34, MS-ESS3-34

What is your Watershed? Earth c 1ABE 3AB 4A 13A; HS-ESS3-34, MS-ESS3-34

Convection in Liquids and Gases Earth c 1ABCDEF 3AB 4A 10B; HS-ESS2-3

Human Dependence and Influence on Oceans Earth c 1ABE 3AB 4AC 10ABC 12E; HS-ESS3-34, MS-ESS2-456

The Great Pacific Garbage Patch Earth c 1ABE 3AB 4AC 10ABC 12E; HS-ESS3-34, MS-ESS2-6, 3-34

Virtual Investigations that go with Human Impacts on Hydrosphere Earth c 1ABEFGH 2ABCD 3ABC 4ABC 11FG 12E; HS-ESS3-34, MS-ESS3-34

Influences on Climate Change Concept Map Earth c 12BCEF; HS-ESS3-56, MS-ESS3-345

Composition of the Atmosphere Earth c 1ABE 3B 4A 6CE 11E; HS-ESS3-6, MS-ESS1-3

The Greenhouse Effect on Earth Earth c 1ABCDEFGH 2BCD 3AB 4A 11C 12CD; HS-ESS2-2, MS-ESS3-5

Climate and Greenhouse Gases: Data Table Earth c 1ABBEFH 2B 3AB 4A 6C 11EF 12CEF; HS-ESS3-6, MS-ESS3-345

Carbon Dioxide and Population Earth c 1ABEF 2B 3AB 4A 11E 12CEF; HS-ESS3-6, MS-ESS3-345

Carbon Emissions Earth c 1ABEF 2BC 3AB 4A 13A; HS-ESS3-256, MS-ESS3-1345

Climate Change Earth c 1ABE 3AB 4A 8H 10C 11ABCD 13B 12CDE; HS-ESS3-56, MS-ESS3-5

Climate Change Poster Earth c 1ABE 4A 12CDE 13A; HS-ESS3-56

Environmental Issues and Ethics Earth c 13A; HS-ESS3-24

Virtual Investigations that go with Influences of Climate Change Earth c 1AEFGH 2ABCD 3ABC 4ABC 11ABCDFG 12FD; HS-ESS2-256, 3-56, MS-ESS3-345

Equipment List for all Investigations

If you want to be able to do all the labs in this manual, here is the list of all the equipment you will need in order of appearance:

- Small Lego Sets
- Water
- Scales
- Meter sticks
- Temperature probes
- Interfaces
- Computers
- Logger Pro
- Graduated cylinders
- Stopwatches
- Rulers
- Pennies
- Rolls of pennies
- Empty penny rolls
- Round balloons
- Internet and textbooks
- Golf balls
- Beads
- Variety of balls
- Embroidery hoops
- Elastic fabric
- Marbles
- String
- Tennis balls
- Wire coat hangers
- Play-Doh
- Colored pencils
- Butcher paper
- Plastic shoebox tubs & lids
- Black paint
- Press'n Seal
- Incandescent work lamps
- Molecular model kits
- Periodic Table
- Ring stand and clamps
- Globes
- Tape
- Soccer balls/Basketballs
- White paper
- Drinking glasses/clear plastic cups
- Ping pong balls
- Knives
- Soft boiled eggs
- Paper towels
- Avocados
- Starbursts

Scissors	Geologic maps
Ziploc bags	Ice
Aluminum foil	Salt water
Beaker tongs	Humidity probes
Hotplates	UVB Sensors
Safety goggles	Shoelaces
Tablespoons	Glue
Soil	Leaves
Beakers	Shelled nuts/seeds
Pipettes	Food serving gloves
Chalk	Large mixing bowl
Sugar cubes	Paper plates
Small trays/pans	Cheese-flavored Goldfish Crackers
Aprons	Pretzel-flavored Goldfish Crackers
Shovels	Soil moisture probes
Large tubs/pans	Heat lamp
Pitchers	Colored construction paper
Spray bottles	Sand
Distilled water	Coffee filters
Stirring rods	Rubber bands
Spatulas	Mulch
Conductivity probes	Sliced white bread
Graham crackers	Food coloring
Frosting	Well-plates
Fruit Roll-ups	Vinegar
Topographic maps	pH probes

2-liter bottles

Turbidity sensors

Lens paper

www.ingramcontent.com/pod-product-compliance
Lightning Source LLC
Chambersburg PA
CBHW080523240526
45472CB00021BA/1741